U.S. Marine Corps Close Combat

Plus

U.S. Marine Corps Afghanistan: An Introduction to the Country and People

Department of Defense

Pentagon
Publishing

DEPARTMENT OF THE NAVY
Headquarters United States Marine Corps
Washington, D.C. 20380-1775

18 February 1999

FOREWORD

1. PURPOSE

Today's Marines operate within a continuum of force where conflict may change from low intensity to high intensity over a matter of hours. Marines are also engaged in many military operations other than war, such as peacekeeping missions or noncombatant evacuation operations, where deadly force may not be authorized. During noncombative engagements, Marines must determine if a situation warrants applying deadly force. Sometimes Marines must decide in a matter of seconds because their lives or the lives of others depend on their actions. To make the right decision, Marines must understand both the lethal and nonlethal close combat techniques needed to handle the situation responsibly without escalating the violence unnecessarily. Marine Corps Reference Publication (MCRP) 3-02B, *Close Combat,* provides the tactics, techniques, and procedures of Marine Corps close combat. It also provides the doctrinal basis for the Marine Corps Close Combat Training Program (MCCCTP).

2. SCOPE

This publication guides individual Marines, unit leaders, and close combat instructors in the proper tactics, techniques, and procedures for close combat. MCRP 3-02B is not intended to replace supervision by appropriate unit leaders and close combat instruction by qualified instructors. Its role is to ensure standardization and execution of tactics, techniques, and procedures throughout the Marine Corps.

3. SUPERSESSION

MCRP 3-02B supersedes Fleet Marine Force Manual (FMFM) 0-7, *Close Combat,* dated 9 July 1993. There are significant differences between the two publications. MCRP 3-02B should be reviewed in its entirety.

4. WARNING

Techniques described in this publication can cause serious injury or death. Practical application in the training of these techniques will be conducted in strict accordance with approved Entry Level Close Combat, Close Combat Instructor (CCI), and Close Combat Instructor Trainer (CCIT) lesson plans. Where serious danger exists, the reac er is alerted by the following:

WARNING

5. CERTIFICATION

Reviewed and approved this date.

BY DIRECTION OF THE COMMANDANT OF THE MARINE CORPS

J. E. RHODES
Lieutenant General, U.S. Marine Corps
Commanding General
Marine Corps Combat Development Command

DISTRIBUTION: 14400006600

OVERVIEW OF CLOSE COMBAT

1. Purpose of Close Combat

Close combat is the physical confrontation between two or more opponents. It involves armed and unarmed and lethal and nonlethal fighting techniques that range from enforced compliance to deadly force. The purpose of close combat is to execute armed and unarmed techniques to produce both lethal and nonlethal results. Unarmed techniques include hand-to-hand combat and defense against hand-held weapons. Armed techniques include techniques applied with a rifle, bayonet, knife, baton, or any weapon of opportunity.

2. Continuum of Force

Marines will find themselves in both combative and noncombative situations. The threat level in these situations can rise and fall several times based on the actions of both Marines and the people involved. The escalation of force stops when one person complies with the demands imposed by another person. This range of actions is known as a continuum of force. Continuum of force is the concept that there is a wide range of possible actions, ranging from voice commands to application of deadly force, that may be used to gain and maintain control of a potentially dangerous situation (MCO 5500.6_, *Arming of Security and Law Enforcement [EE] Personnel and the Use of Force)*. Continuum of force consists of five levels that correspond to the behavior of the people involved and the actions Marines use to handle the situation (see the table below). Close combat techniques are executed in levels three, four, and five.

Level One: Compliant(Cooperative)

The subject complies with verbal commands. Close combat techniques do not apply.

Level Two: Resistant (Passive)

The subject resists verbal commands but complies immediately to any contact controls. Close combat techniques do not apply.

Continuum of Force		
Level	**Description**	**Actions**
1	Compliant (Cooperative)	Verbal commands
2	Resistant (Passive)	Contact controls
3	Resistant (Active)	Compliance techniques
4	Assaultive (Bodily Harm)	Defensive tactics
5	Assaultive (Serious Bodily Harm/ Death)	Deadly force
Note: Shading indicates levels in which Marines use close combat techniques.		

Level Three: Resistant (Active)

The subject initially demonstrates physical resistance. Marines use compliance techniques to control the situation. Level three incorporates close combat techniques to physically force a subject to comply. Techniques include—

- Come-along holds.
- Soft-handed stunning blows.
- Pain compliance through joint manipulation and the use of pressure points.

Level Four: Assaultive (Bodily Harm)

The subject may physically attack Marines, but he does not use a weapon. Marines use defensive tactics to neutralize the threat. Defensive tactics include the following close combat techniques:

- Blocks.
- Strikes.
- Kicks.
- Enhanced pain compliance procedures.
- Nightstick blocks and blows.

Level Five: Assaultive (Serious Bodily Harm/Death)

The subject usually has a weapon and will either kill or seriously injure someone if he is not stopped immediately and brought under control. Typically, to control the subject, Marines apply deadly force through the use of a firearm, but they may also use armed and unarmed close combat techniques.

3. Marine Corps Tactical Concepts

Close combat techniques support the following key Marine Corps tactical concepts. The concepts are not standalone ideas but are to be combined to achieve an effect that is greater than their separate sum.

Achieving a Decision

Achieving a decision is important in close combat. An indecisive fight wastes energy and possibly Marines' lives. Whether the intent is to control an opponent through restraint or defend themselves in war, Marines must have a clear purpose before engaging in close combat and act decisively once engaged.

Gaining an Advantage

A basic principle of martial arts is to use the opponent' s strength and momentum against him to gain more leverage than one's own muscles alone can generate, thereby gaining an advantage. In close combat, Marines must exploit every advantage over an opponent to ensure a successful outcome. This can include employing various weapons and close combat techniques that will present a dilemma to an opponent. Achieving surprise can also greatly increase leverage. Marines try to achieve surprise through deception, stealth, and ambiguity.

Speed

Marines use speed to gain the initiative and advantage over the enemy. In close combat, the speed and violence of the attack against an opponent provides Marines with a distinct advantage. Marines must know and understand the basics of close combat so they can act instinctively with speed to execute close combat techniques.

Adapting

Close combat can be characterized by friction, uncertainty, disorder, and rapid change. Each situation is a unique combination of shifting factors that cannot be controlled with precision or certainty. For example, a crowd control mission may call for Marines to employ various techniques ranging from nonlethal restraint to more forceful applications. Marines who adapt quickly will have a significant advantage.

Exploiting Success

Typically, an enemy will not normally surrender simply because he was placed at a disadvantage. Marines cannot be satisfied with gaining an advantage in a close combat situation. They must exploit any advantage aggressively and ruthlessly until an opportunity arises to completely dominate the opponent. Marines must exploit success by using every advantage that can be gained.

(reverse blank)

Close Combat

Table of Contents

CHAPTER 1

FUNDAMENTALS OF CLOSE COMBAT

This chapter describes all techniques for a right-handed person. However, all techniques can be executed from eittwr side.

The Marine is depicted in camouflage utilities. The opponent is depicted without camouflage.

The fundamentals of close combat include ranges, weapons of the body, target areas of the body, and pressure points of the body. These fundamentals form the basis for all close combat techniques. They provide Marines with a common framework regardless of the type of confrontation or the techniques used. If Marines apply these fundamentals properly in a close combat situation, they may save their lives or the lives of fellow Marines.

1. Ranges of Close Combat

Close combat engagements occur within three ranges: long range, midrange, and close range. During any engagement, these ranges may blur together or may rapidly transition from one to another until either the opponent is defeated or the conflict is resolved.

Long Range

During long range engagements, combatants engage each other with rifles, bayonets, sticks, or entrenching tools. See figure below.

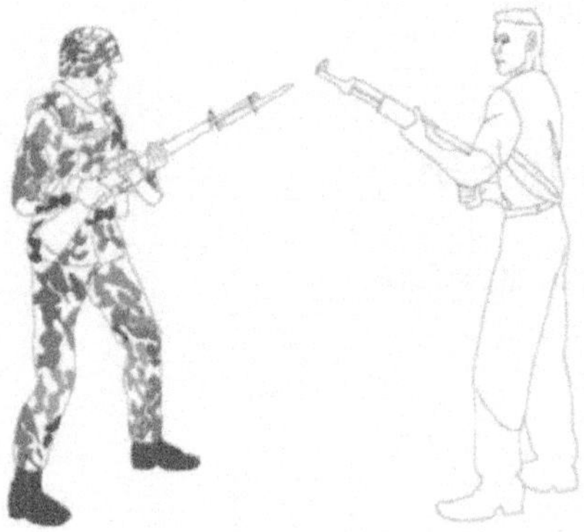

Midrange

During midrange engagements, combatants engage each other with knives, punches, or kicks.

Close Range

During close range engagements, combatants grab each other. Close range engagements also involve elbow strikes, knee strikes, and grappling.

2. Weapons of the Body

Hands and Arms

The hands, forearms, and elbows are the arm's individual weapons. The hands consist of several areas that can be used as weapons: fists, edges of hands, palms, and fingers.

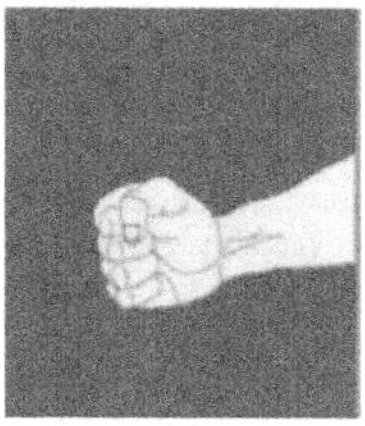

Fists. To minimize injury to the fists, Marines use their fists as weapons to target soft tissue areas such as the throat. The fists' striking surfaces are the first two knuckles of the hands or the meaty portions of the hands below the little fingers.

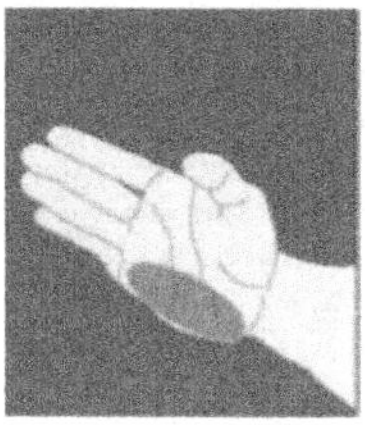

Edge of Hand. Marines use the edge of the hand (knife edge) as a weapon. Marines use the edge of the hand to strike soft tissue areas.

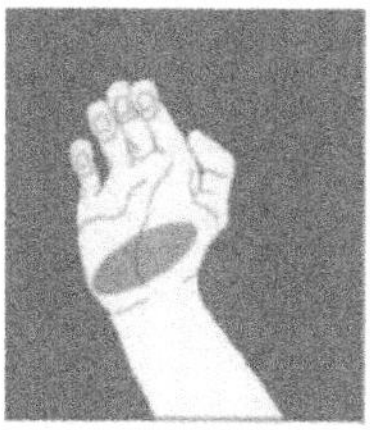

Palms. Because of the palm's padding, Marines use the heels of the palms to strike, parry, and/or block.

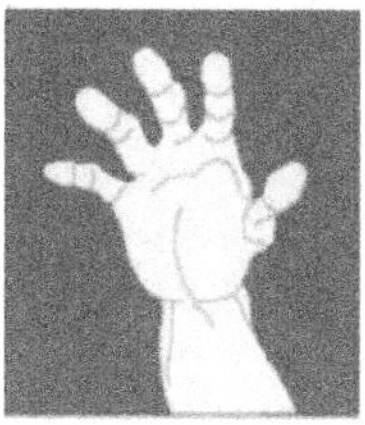

Fingers. Marines use the fingers to gouge, rip, and tear soft tissue areas (e.g., eyes, throat, groin).

Forearms. Marines use the forearms as a defensive tool to deflect or block attacks. Forearms can also be used as striking weapons to damage or break an opponent's joints and limbs. Marines sustain less self-injury when strikes are conducted with the forearms than when strikes are conducted with fists and fingers.

Elbows. Marines use the elbows as striking weapons. Because of the short distance needed to generate power, elbows are excellent weapons for striking during the close range of close combat.

Legs

The legs are more powerful than any other weapon of the body, and they are less prone to injury when striking. The feet are protected by boots and are the preferred choice for striking.

Feet. Marines use the balls of the feet, the insteps, and the toes to kick an opponent. Marines use the cutting edge of the heels and the heels to stomp on an opponent. Marines must be wearing boots when striking with the toes.

Knees. Like elbows, knees are excellent weapons in the close range of close combat. Knee strikes are most effective while fighting close to an opponent where kicks arc impractical. The opponent's groin area is an ideal target for the knee strike if he is standing upright. Knee strikes can deliver a devastating secondary attack to an opponent's face following an initial attack that caused him to bend at the waist.

3. Target Areas of the Body

During close combat, Marines strive to attack the accessible target areas of an opponent's body. The readily accessible areas will vary with each situation and throughout the engagement. The target areas are divided into five major groups: head, neck, torso, groin, and extremities. The figure below illustrates target areas of the body.

Head

The vulnerable regions of the head are the eyes, temple, nose, ears, and jaw. Massive damage to the head kills an opponent.

Eyes. The eyes are excellent targets because they are soft tissue areas that arc not protected by bone or muscle. Attacks to this area may cause the opponent to protect the area with his hands, allowing Marines to execute a secondary attack to other

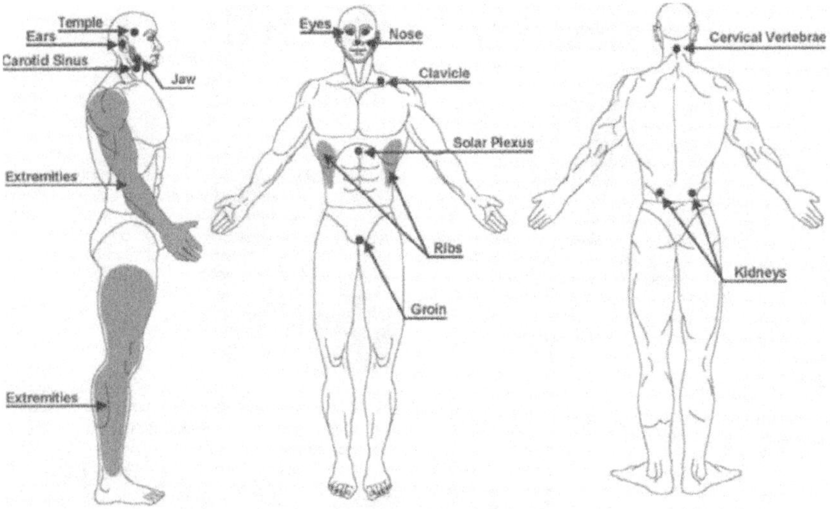

target areas while the opponent uses his hands to protect his eyes.

Temple. The temple is one of the most fragile areas of the head. Powerful strikes to the opponent's temple cause permanent damage and death.

Nose. The nose is very sensitive and easily broken. An attack to this area causes involuntary watering and closing of the opponent's eyes, rendering him vulnerable to secondary attacks. However, through training, individuals can condition themselves to withstand attacks to the nose. Therefore, any attack to the nose must be powerfully delivered.

Ears. Attacks to the ears may cause the eardrum to rupture. But this may not stop or even distract an opponent unless Marines powerfully deliver the strike.

Jaw. The jaw region, when struck forcefully, renders the opponent unconscious. Strikes to the jaw cause painful injuries to the teeth and surrounding tissues (e.g., lips, tongue), but the risk of self-injury is great unless Marines deliver strikes with a hard object such as a helmet, rifle butt, or boot heel.

Neck

The front of the neck, or throat area, is a soft tissue area that is not covered by natural protection. Damage to this region causes the opponent's trachea to swell, closing his airway, which can lead to death.

Carotid Sinus. The carotid sinus is located on both sides of the neck just below the jaw. Strikes to the carotid sinus restrict blood flow to the brain, causing loss of consciousness or death.

Cervical Vertebrae. The cervical vertebrae on the back of the neck, from the base of the skull to the top of the shoulders, contains the spinal cord, which is the nervous system's link to the brain. The weight of the head and the lack of large muscle mass allow damage to the cervical vertebrae and spinal cord. Excessive damage to this area causes pain, paralysis, or death.

Torso

Clavicle. The opponent's clavicle (or collar bone) can be easily fractured, causing immobilization of the arm.

Solar Plexus. Attacks to the opponent's solar plexus or center of the chest can knock the breath out of him and immobilize him.

Ribs. Damage to the opponent's ribs immobilizes him. It may also cause internal trauma.

Kidneys. Powerful attacks to the opponent's kidneys cause immobilization, permanent damage, or death.

Groin

The groin area is another soft tissue area not covered by natural protection. Any damage to this area causes the opponent to involuntarily protect his injured area, usually with his hands or legs. In male opponents, the scrotum is the main target since even a near miss causes severe pain, contraction of the lower abdominal muscles, deterioration of his stance, and possible internal trauma.

Extremities

Rarely will an attack to the opponent's extremities (arms and legs) cause death, but they are important target areas in close combat. Damage to an opponent's joints causes immobilization.

4. Pressure Points of the Body

There are nerves in the human body that, when pressure is applied or when they are struck, allow Marines to control a subject through pain compliance. Marines use pressure points to control an opponent when deadly force is not authorized. They also use pressure points to soften or distract an opponent so a lethal or nonlethal technique can be employed. The figure on page 1-5 illustrates the body's pressure points. Marines execute attacks to pressure points by—

: Rapidly kicking or striking pressure points.
: Slowly applying steady pressure to pressure points.

Infraorbital Nerve

The infraorbital nerve is just below the nose. Marines apply pressure to this nerve with an index finger to control the opponent.

Mastoid Process

The mastoid process is behind the base of the ear and beneath the edge of the jaw. Marines apply inward and upward pressure to this pressure point with the fingers to distract and control the opponent.

Jugular Notch

The jugular notch is at the base of the neck in the notch formed at the center of the clavicle. Marines apply pressure in a quick, stabbing motion with the index finger. Strikes to the jugular notch cause serious damage.

Brachial Plexus (Tie In)

The brachial plexus (tie in) is on the front of the shoulder at the joint. Strikes and pressure applied with the hand are effective on this nerve.

Radial Nerves

Radial nerves are on the inside of the forearms along the radius bones. Strikes and pressure applied with the hand to the radial nerve serve as a softening technique.

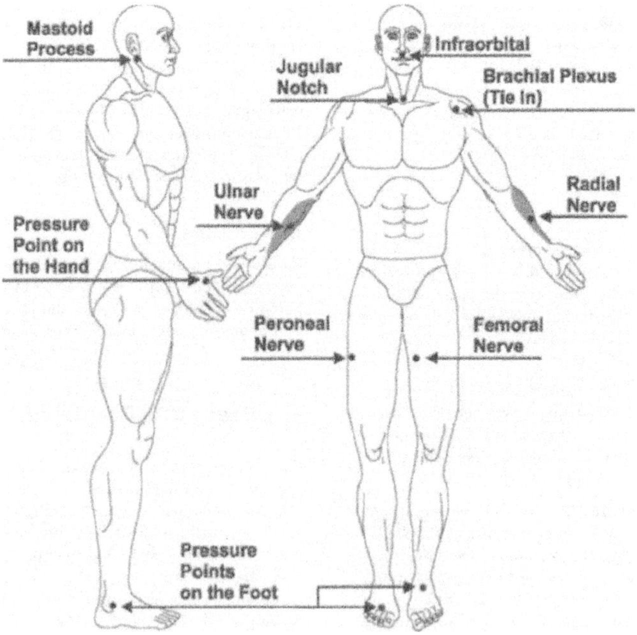

Ulnar Nerve

Ulnar nerves are on the outside of the forearms along the ulnar bones. Strikes and pressure applied with the hand to the ulnar nerve serve as a softening technique.

Pressure Point on the Hand

The hands contain a pressure point on the webbing between the thumbs and index fingers where the two bones of the fingers meet.

To force an opponent to soften or release his grip, Marines apply pressure with their index fingers to this pressure point or strike this pressure point with their fists.

Femoral Nerves

Femoral nerves are on the inside of the thighs along the femur bones. Strikes to the femoral nerve serve as a softening technique.

Peroneal Nerves

Peroneal nerves are on the outside of the thighs along the femur bones. Strikes to the peroneal nerve serve as a softening technique.

Pressure Points on the Feet

There are pressure points on the feet that, when pressure is applied or when they are struck, serve to soften or distract the opponent. Marines apply pressure with the toe, edge, or heel of their boots to the following points:

: The notch below the ball of the ankle.
: The top center of the foot, above the toes.
: The top of the foot where the leg and foot meet.

5. Basic Warrior Stance

Achieving a solid stance is critical to stability and movement throughout any close combat confrontation. The basic warrior stance provides the foundation for stability and movement that is needed to execute close combat techniques. To execute the basic warrior stance, Marines put their feet apart, hands up, elbows in, and chin down.

Feet Apart

Place feet shoulder-width apart.

Keep the head forward and eyes on the opponent, take a half step forward with the left foot, and pivot on the heels so the hips and shoulders are at approximately a 45-degree angle to the right.

Distribute body weight evenly on both legs. Bend the knees slightly.

Hands Up

Curl the fingers naturally into the palm of the hand. Position the thumb across the index and middle fingers. Do not clench the fists. Clenching the fists increases muscular tension in the forearms and decreases speed and reaction time.

Bring the hands up to the face at chin level, with the palms facing each other. Hold the fists up high enough to protect the head, but not so high that they block the field of vision. Ensure continuous eye contact with the opponent.

Elbows In

Tuck the elbows in close to protect the body.

Chin Down

Tuck the chin down to take advantage of the natural protection provided by the shoulders.

6. Angles of Approach and Movement

Marines use movement to control a confrontation and to retain a tactical advantage. Movement increases power and maximizes momentum. By moving around the opponent, Marines gain access to different target areas of the opponent's body. Movement allows Marines to use different weapons of the body and different close combat techniques to attack specific target areas.

Angles of Approach

Marines move anywhere within a 360-degree circle around the opponent to gain a tactical advantage. This circle provides access to different target areas of the opponent's body.

When facing an opponent, Marines move in a 45-degree angle to either side of the opponent. Moving at a 45-degree angle avoids an opponent's strike and puts Marines in the best position to attack the opponent. Marines should avoid being directly in front of an opponent during a confrontation. If a Marine is directly in front of an opponent, the opponent can rely on his forward momentum and linear power to seize the tactical advantage.

Movement

Marines must know how to move in all directions while maintaining the basic warrior stance. During any movement, the legs or feet should not be crossed. Once a movement is completed, the basic warrior stance should be resumed. Maintaining the basic warrior stance protects Marines and puts them in the proper position to launch an attack against an opponent.

Note: Before body movement begins, Marines turn their heads quickly to the new direction. The faster the head turns, the faster the body moves, and the quicker Marines attain visual contact with the opponent.

Forward to the Left. To move forward to the left from the basic warrior stance, Marines-

: Move the left foot forward at a 45-degree angle from the body (approximately 12 to 15 inches), keeping the toe pointed toward the opponent.

Bring the right foot behind the left foot as soon as the left foot is in place. This returns Marines to the basic warrior stance.

Forward to the Right. To move forward to the right from the basic warrior stance, Marines-

Move the right foot forward at a 45-degree angle from the body (approximately 12 to 15 inches).

Bring the left foot, toe pointing toward the opponent, in front of the right foot as soon as the right foot is in place. This returns Marines to the basic warrior stance.

Backward to the Left. To move backward to the left from the basic warrior stance, Marines execute the forward movement in reverse. Marines-

Move the left foot backward at a 45-degree angle from the body (approximately 12 to 15 inches), keeping the toe pointed toward the opponent.

Bring the right foot behind the left foot as soon as the left foot is in place. This returns Marines to the basic warrior stance.

Backward to the Right. To move backward to the right from the basic warrior stance, Marines execute the forward movement in reverse. Marines-

Move the right foot backward at a 45-degree angle from the body (approximately 12 to 15 inches).

Bring the left foot, toe pointing toward the opponent, in front of the right foot as soon as the right foot is in place. This returns Marines to the basic warrior stance.

7. Balance and Off-Balancing

Balance

In any close combat situation, Marines must strive to maintain balance. The last place to be in a close combat situation is on the ground. Marines must maintain a strong base and a low center of balance, their feet must be a shoulder-width apart, and they must stay on their toes to enable quick movement.

Off-balancing

Marines use off-balancing techniques to control an opponent. These techniques are used to throw an opponent to the ground while Marines remain standing, or they are used to put Marines in a position for an offensive attack.

Off-balancing techniques use the opponent's momentum to move or throw him. For example, if the opponent is charging a Marine, the Marine pulls the opponent to drive him to the ground. Likewise, if the opponent is pulling a Marine, the Marine pushes the opponent to drive him to the ground.

Off-balancing techniques also rely on the power generated by the opponent. For example, during combat a Marine may be tired or outnumbered. Depending on the generated energy and momentum of the opponent, the Marine employs off-balancing techniques with very little effort and still provides effective results.

Because off-balancing techniques rely on the momentum and power generated by the opponent, these techniques are particularly effective for Marines who may be outsized by their opponent or lack their opponent's strength.

Angles of Off-balancing

There are eight angles or directions in which an opponent can be off-balanced: forward, rear, right, left, forward right, forward left, rear right, and rear left.

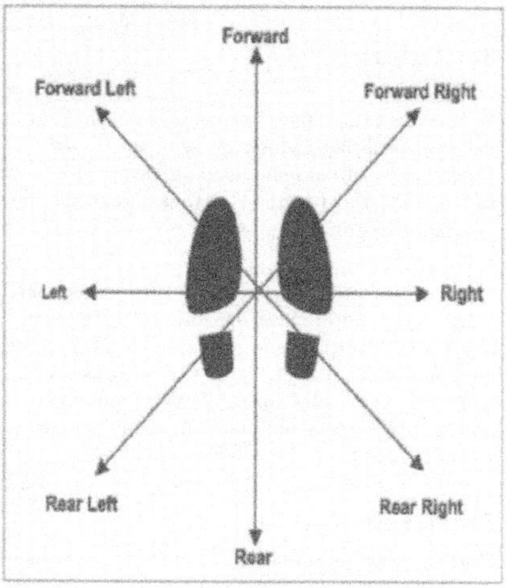

Note: The angles correspond to the Marine's perspective, not the opponent's. Forward, rear, right, and left are straight angles. Forward right, forward left, rear right, and rear left are considered quadrants that are at a 45-degree angle in either direction to the front or rear.

Off-balancing Techniques

Marines off-balance an opponent by pushing, pulling, or bumping the opponent with their hands, arms, or bodies.

To pull, Marines grab an opponent with their hands and drive him forcefully to one of the rear quadrants or to the right or left.

To push, Marines grab the opponent with their hands and drive him forcefully into one of the front quadrants or to the right or left.

Marines execute bumping in the same manner as pushing, but use their shoulders, hips, and legs instead of their hands to off-balance the opponent.

8. Falls

Marines may lose their balance or be thrown to the ground during encounters with an opponent. Marines use falling techniques to absorb the impact of a fall and to quickly return to their feet following an opponent's attack.

Whether falling or being thrown by an opponent, Marines strive to reduce the force of the impact, to prevent serious personal injury, and to increase the chances of survival. Falling techniques use the body's large muscles (back, thighs, buttocks) to protect vital organs and bones from injury and immobilization.

Front Fall

Marines execute a front fall to break a fall on the front. To execute the front fall, Marines-

Bend the elbows and place the palms facing out in a position to spread and absorb the impact of the fall.

Fall forward, breaking the fall with the forearms and palms. The forearms and hands, down to the fingertips, should strike the ground simultaneously.

Offer resistance with the forearms and hands to keep the head raised off the ground.

Side Fall

Marines execute a side fall to break a fall on the side. To execute the side fall, Marines-

Bring the right arm across the body so the hand is next to the left shoulder with the palm facing inboard.

Fall to the side, breaking the fall with the right arm by slapping the ground and making contact from the shoulder or forearm down to the hand. At the same time, tuck the chin and keep the head raised off the ground. The chin should be tucked to the chest at all times to prevent whiplash.

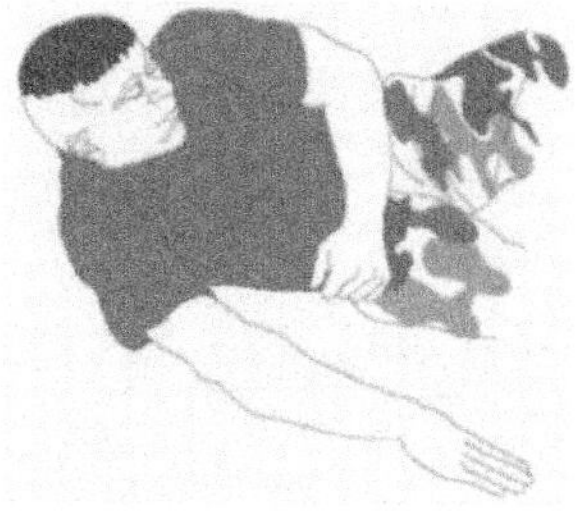

Stretch the right leg out to make contact with the ground and to distribute and absorb the impact. Bend the left leg, allowing the foot to make contact with the ground.

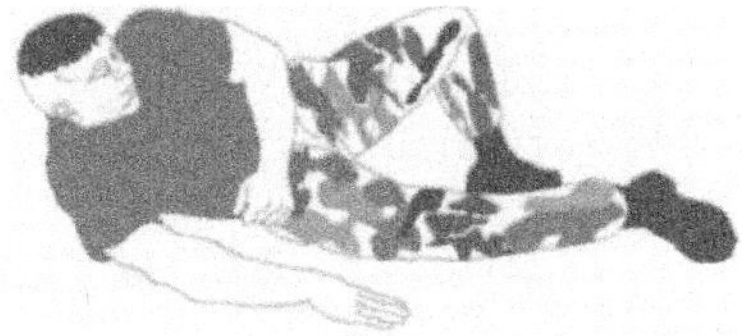

Back Fall

Marines execute a back fall to break the fall when being thrown or falling backward. To execute the backfall, Marines-

Cross the hands in front of the chest and tuck the chin.

Fall backward and slap the ground with the forearms and hands to absorb the impact of the fall and keep the head off the ground.

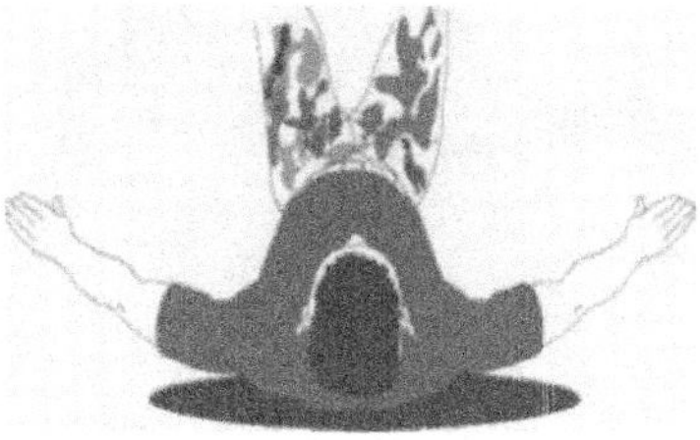

Forward Shoulder Roll

Marines use the forward shoulder roll to break a fall from an opponent's attack and to use the momentum of the fall to get back on their feet quickly. Ideally, Marines execute the forward shoulder roll to a standing position so they can continue fighting. To execute the forward shoulder roll, Marines-

Contact the ground with the back of the right forearm and upper arm. Tuck the chin into the chest.

Roll onto the right shoulder, rolling diagonally across the back to land on the left hip.

Slap the ground with the left arm, absorbing the impact from the shoulder to the hand, palm down.

Keep the left leg straight to absorb as much of the impact as possible. The right leg is bent and the foot hits flat on the ground.

Bend the left leg upon impact to push off with the left knee and leg to a squatting and then a standing position. Forward momentum should carry the Marine to a standing position.

(reverse blank)

CHAPTER 2

LETHAL AND NONLETHAL WEAPONS TECHNIQUES

This chapter describes all techniques for a right-handed person. However, all techniques can be executed from either side.

In drawings, the Marine is depicted in woodland camouflage utilities; the opponent is depicted without camouflage. In photographs, the Marine is depicted in woodland camouflage utilities; the opponent is depicted in desert camouflage utilities.

1. Bayonet Techniques

WARNING

During training, Marines must have bayonets sheathed. Marines use bayonet dummies to practice bayonet techniques. When practicing offensive and defensive bayonet techniques student-on-student, Marines use pugil sticks.

All Marines armed with a rifle carry a bayonet. The bayonet is an effective weapon if Marines are properly trained in offensive and defensive bayonet techniques. An offensive attack, such as a thrust, is a devastating attack that can quickly end a fight. Defensive techniques, such as the block and parry, can deter the opponent's attack and allow Marines to regain the initiative. Through proper training, Marines develop the courage and confidence required to effectively use a bayonet to protect themselves and destroy the enemy. In situations where friendly and enemy troops are closely mingled and rifle fire and grenades are impractical, the bayonet becomes the weapon of choice.

Holding the Rifle

To execute bayonet techniques, Marines hold the rifle in a modified basic warrior stance. All movement begins and ends with the basic warrior stance. To hold the rifle, Marines—

Use an overhanded grasp to grab the small of the rifle's stock. Use an underhanded grasp to grab the handguards of the rifle.

Lock the buttstock of the rifle against the hip with the right forearm.

Orient the blade end of the rifle toward the opponent.

Offensive Bayonet Techniques

Straight Thrust. Marines use the straight thrust to disable or kill an opponent. It is the most deadly offensive technique because it causes the most trauma to an opponent. Target areas are the opponent' s throat, groin, or face. The opponent's chest and stomach are also excellent target areas if not protected by body armor or combat equipment. To execute the straight thrust, Marines—

Lift the left leg and lunge forward off the ball of the right foot while thrusting the blade end of the weapon forward, directly toward the opponent.

Retract the weapon and return to the basic warrior stance.

Slash. Marines use the slash to kill an opponent or to create an opening in his defense. The target area is the opponent's neck. To execute the slash, Marines—

Extend the left hand back toward the left shoulder.

Thrust the left hand forward and swing it to the right, bring the right hand back toward the hip, and turn the cutting edge of the blade toward the opponent's neck. The movement is a slashing motion so the blade cuts across the opponent's neck.

Horizontal Buttstroke. Marines use the horizontal buttstroke to weaken an opponent's defenses, to cause serious injury, or to set him up for a killing blow. Target areas are the opponent's head, neck, and legs. To execute the horizontal buttstroke, Marines—

Step forward with the right foot and drive the right hand forward. Rotate the hips and shoulders into the strike. Move the left hand back toward the left shoulder.

Strike the opponent with the butt of the weapon.

Vertical Buttstroke. Marines use the vertical buttstroke to weaken an opponent's defenses, to cause serious injury, or to set him up for a killing blow. Target areas are the opponent's groin and face. To execute the vertical buttstroke, Marines—

Step forward with the right foot and drive the right hand straight up.

Pull the left hand back over the left shoulder.

Strike the opponent with the butt of the weapon.

Smash. Marines use the smash as a follow-up technique to the horizontal or vertical buttstroke, particularly if they missed the target. The target area is the opponent's head. To execute the smash following a buttstroke, Marines-

Step forward with the right foot and place the blade end of the weapon over the left shoulder and elevate the right elbow above the shoulders.

Drive the arms straight forward, striking the opponent with the butt of the weapon.

Defensive Bayonet Techniques

Parry. Marines use a parry as a defensive technique to redirect or deflect an attack. A parry is a slight redirection of a linear attack by an opponent; e.g., a straight thrust or a smash. To execute the parry, Marines-

Use the bayonet end of the rifle to redirect the barrel or bayonet of the opponent's weapon.

Lock the weapon against the hip with the right forearm.

Rotate to the right or left, moving the bayonet end of the rifle to parry the opponent's attack. Rotation is generated from the hips.

Redirect or guide the opponent's weapon away from the body by exerting pressure against the opponent' s weapon.

High Block. Marines execute a high block against a vertical attack coming from high to low. To execute the high block, Marines—

Thrust the arms up forcefully at approximately a 45-degree angle from the body. The weapon should be over the top of the head and parallel to the ground. The elbows are bent, but there is enough muscular tension in the arms to absorb the impact and deter the attack.

Low Block. Marines execute the low block against a vertical attack coming from low to high. To execute the low block, Marines—

Thrust the arms down forcefully at approximately a 45-degree angle from the body. The weapon should be at or below the waist and parallel to the ground. The elbows are bent, but there is enough muscular tension in the arms to absorb the impact and deter the attack.

Left and Right Block. Marines execute a left or right block against a horizontal buttstroke or a slash. To execute the left or right block, Marines-

Thrust the arms forcefully to the right or left, holding the rifle vertically in the direction of the attack. The elbows are bent, but there is enough muscular tension in the arms to absorb the impact and deter the attack.

Counter Action Following the Block. After deflecting an opponent's attack with a block, Marines counter with a slash or a horizontal buttstroke to regain the initiative. However, the objective in any bayonet fight is to thrust forward with the blade end of the weapon to immediately end the fight.

Group Strategy

On occasion, Marines may engage an opponent as a member of a group or numerous opponents by one's self or as a member of a group. By combining bayonet fighting movements and simple strategies, Marines can effectively overcome their opponent or opponents.

Offensive Strategy: Two Against One. If two bayonet fighters engage one opponent, the fighters advance together.

Fighter 1 engages the opponent while fighter 2 swiftly and aggressively attacks the opponent's exposed flank and destroys the opponent.

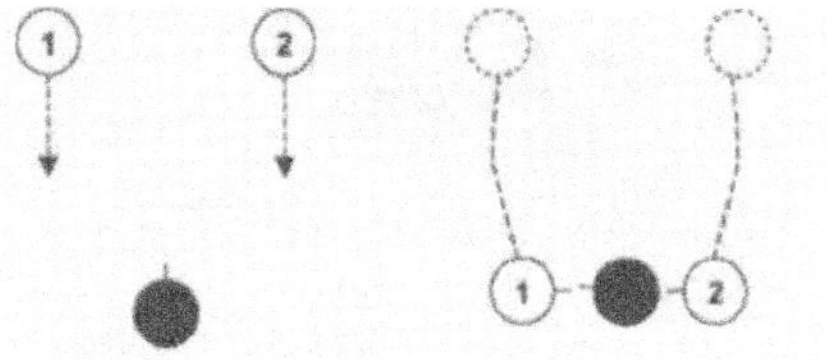

Offensive Strategy: Three Against Two. If three bayonet fighters engage two opponents, the fighters advance together keeping their opponents to the inside.

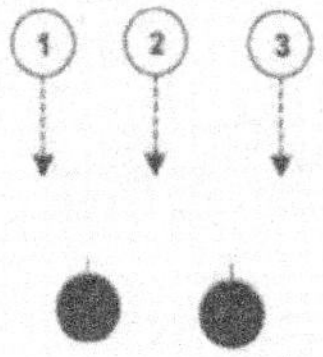

Fighters 1 and 3 engage opponents. Fighter 2 attacks the opponent's exposed flank engaged by fighter 1 and destroys the opponent.

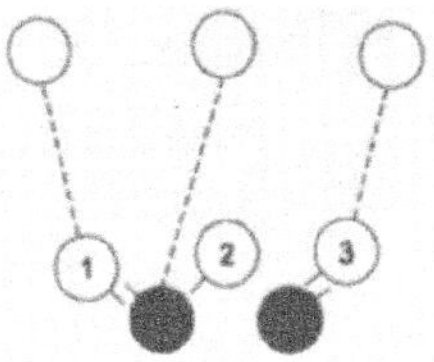

Fighters 1 and 2 turn and attack the exposed flank of the opponent engaged by fighter 3 and destroy the opponent.

Defensive Strategy: One Against Two. If a fighter is attacked by two opponents, the fighter immediately positions himself at the flank of the nearest opponent and keeps that opponent between himself and the other opponent.

Using the first opponent's body as a shield against the second opponent, the fighter destroys the first opponent quickly before the second opponent moves to assist.

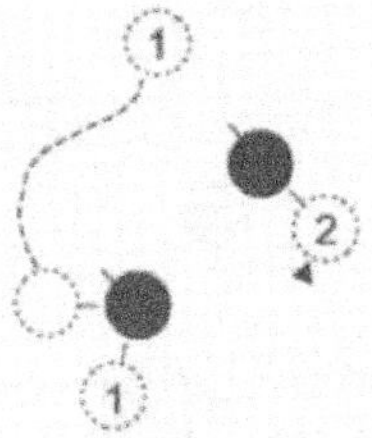

Then, the fighter engages and destroys the second opponent.

Defensive Strategy: Two Against Three. If two fighters are attacked by three opponents, the fighters immediately move to the opponent's flanks.

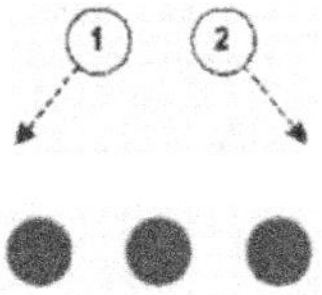

Fighters 1 and 2 quickly attack and destroy their opponents before the third opponent closes in.

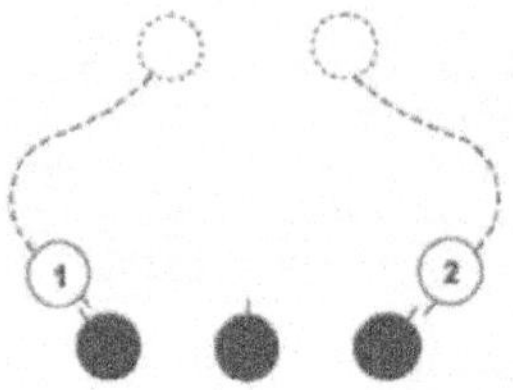

Fighter 1 engages the third opponent while fighter 2 attacks the opponent's exposed flank and destroys the opponent.

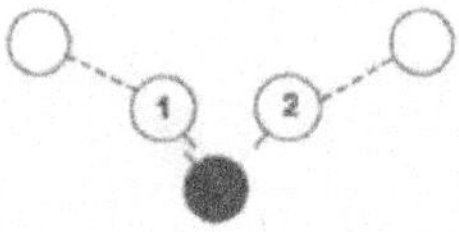

2. Nonlethal Rifle and Shotgun Retention Techniques

Most Marines are armed with the M16A2 service rifle. Marines are taught to keep their weapon with them at all times. Marines must be constantly alert to their surroundings and the people moving in and around their environment. Marines may be confronted by an individual who tries to take theii weapons. If this happens, Marines should not struggle with the individual. To retain positive control of their weapons, Marines must understand and apply weapons retention techniques, otherwise known as armed manipulation. The fol lowing techniques can be used with either the rifle or the shotgun.

Blocking Technique

To execute a blocking technique, Marines-

Stand in a defensive posture.

Use the weapon to block the opponent by thrusting it out firmly, with the elbows bent. Do not try to hit the opponent with the rifle. The rifle is used as a barrier.

Technique if Opponent Grabs Weapon Underhanded

Usually an opponent will try to grab the weapon or block it as an instinctive action. If the opponent uses an underhand grab to seize the handguards of the weapon, Marines—

Trap the opponent's closest finger(s) above the knuckle with the thumb so he cannot release his grip-

Apply bone pressure on the opponent's finger to initiate pain compliance.

Rotate the barrel of the weapon up or down quickly while maintaining pressure on the opponent's hand. At the same time, quickly pivot to off-balance the opponent.

Technique if Opponent Grabs Weapon Overhanded

If the opponent uses an overhand grab to seize the weapon overhanded, Marines—

Trap the opponent's finger to hold his hand in place.

Rotate the barrel to place it across the opponent's forearm and apply downward pressure. This ac-

tion is similar to an armbar.

Technique if the Opponent Grabs the Muzzle

If the opponent grabs the muzzle of the weapon, Marines—

Rotate the muzzle quickly in a circle motion, Slash downward with the muzzle to release the opponent's grip.

Butt Strikes

Strikes with the butt of the weapon control or ward off an attacker. During any of the retention techniques, Marines use the heel or cutting edge of the weapon to deliver butt strikes to the inside or the outside of the opponent's thigh. The inside butt strike targets the femoral nerve as illustrated in the photo to the left.

The outside butt strike targets the peroneal nerve as illustrated in the photo to the left. Strikes can be made to the outside or inside of the thighs. If a strike to one side of the thigh misses, Marines follow back through with the butt of the weapon on the other side of the thigh.

Off-balancing Techniques

Marines apply off-balancing techniques to throw an opponent to the ground and retain possession of the weapon.

If the opponent grabs the weapon and pushes, Marines should not push on the weapon. They should—

Move with the momentum and movement of the

opponent by pivoting in the direction of the movement by stepping back.

Throw the opponent to the ground with a quick jerking movement by lowering the mu/./le and swinging the butt of the weapon.

If the opponent grabs the weapon and pulls, Marines—

Step on the opponent's foot and push forward to off-balance him and drive him to the ground.

Sweep the opponent's feet out from under him by hooking his leg with the leg and kicking backward.

3. Nonlethal Handgun Retention Techniques

Many Marines are armed with the M9 service pistol. Marines must keep their weapons in their possession at all times. Marines must be constantly alert to their surroundings and the people moving in and around the environment. Marines may be confronted by an individual who tries to take their weapons. To retain positive control of the weapon, Marines must understand and apply handgun retention techniques.

Blocking Technique

Marines perform the following blocking techniques if an opponent attempts to grab their pistol in the holster. Marines—

Place the body between the weapon and the opponent by immediately pivoting so the weapon is away from the opponent.

Step back and away from the opponent while placing the hand on the pistol grip.

Extend the left hand and block, deflect, or strike the opponent's arm.

Armbar Technique

Marines use the armbar technique when an opponent uses his right hand to grab their pistol in the holster. To execute the armbar technique, Marines—

Trap the opponent's right hand by grasping the opponent's wrist or hand with the right hand and applying pressure against the body.

Step back with the right foot and pivot sharply to the right to be next to the opponent. Always pivot in a direction that keeps the weapon away from the opponent.

Straighten the opponent's arm to apply an armbar. The arm should be straight across the torso.

Continue pivoting to the right while pulling back on the opponent's shoulder. This action may break the opponent's arm.

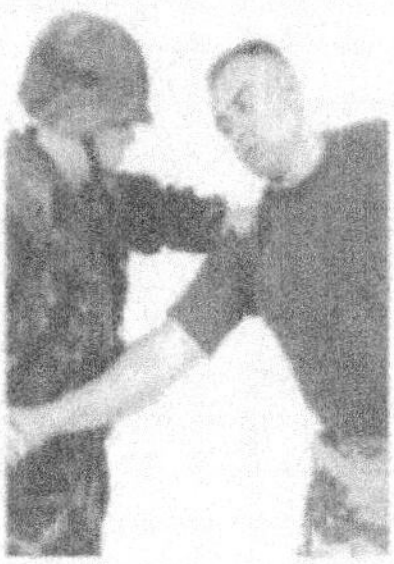

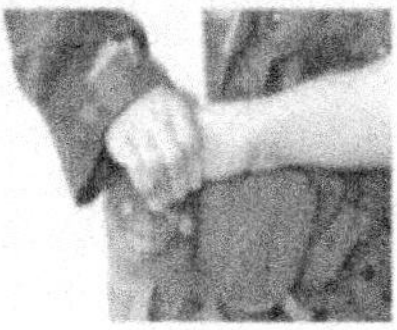

Place the left hand across the opponent's face and apply pressure back and down to take the oppo-

nent to the ground. Pressure applied beneath the nose or on the trachea is equally effective.

Wristlock Technique

Marines use the wristlock technique when an opponent grabs their pistol while it is in the holster with his right hand. To execute the w ristlock technique, Marines—

Grasp the opponent's wrist or hand with the right hand and apply pressure against the body.

Step back and away from the opponent and pivot to the right so the weapon is away from the opponent. Always pivot in a direction that keeps the weapon away from the opponent.

Reach over the opponent's arm with the left hand and grab his hand, applying pressure against his arm with the left forearm. Execute a wristlock.

Incorporate the second hand into the wristlock and, stepping back with the left foot, pivot to the left.

Execute a two-handed wristlock by exerting downward pressure with the thumb and rotating his hand to the left.

Continue to pivot to off-balance the opponent and drive him to the ground.

Note: If the opponent grabs the pistol with his left hand, Marines execute the wristlock with one hand and step in toward the opponent, rather than

away from the opponent.

Softening Techniques

Handgun retention techniques use softening techniques applied to pressure points. Bone pressure

and strikes with the hands (i.e., hammer fist), knees, and feet are also effective softening techniques.

Pressure Points. Marines use pressure point techniques to get the opponent to loosen his grip. Marines use their finger tips to apply pressure to the webbing between the index finger and thumb, the jugular notch, and the brachial plexus tie in. The following figure illustrates pressure applied to the brachial plexus tie in.

Bone Pressure. Bone pressure is the application of pressure on a bone against a hard object to initiate pain compliance. To apply bone pressure, Marines use their hand to trap the opponent's hand on the weapon. Marines apply a slow, steady pressure to the opponent's hand and fingers until his grip is softened or he releases his hold.

Strikes. If it is difficult to apply a retention technique, Marines employ strikes or kicks to force the opponent to loosen his grip. Strikes to the

eyes, the arms (radial nerve), or shoulder (brachial plexus tic in) soften the opponent's grip on the weapon.

Kicks and knee strikes to the peroneal nerve, the femoral nerve, or the groin arc effective because the opponent is typically unprepared to counter the strike.

Stomping on the top of the opponent's foot may distract him or loosen his grip on the weapon.

4. Firearm Disarmament Techniques

Marines use firearm disarmament techniques during a close-range confrontation if they are un-

armed and the opponent has a firearm (pistol). These techniques are equally effective if Marines are armed but do not have time to withdraw and present the weapon. The goal of firearm disarmament techniques is to gain control of the situation

so Marines gain the tactical advantage. The goal is not necessarily to get control of the opponent's weapon.

Pistol to the Front

This technique is used when Marines are unarmed and the opponent has a pistol pointing at their front (e.g., head, chest). The technique is the same if the opponent sticks the pistol under the Marine's chin. To execute the counter when an opponent is pointing a pistol toward the front of a Marine, Marines—

Place the hands close to the weapon, about chest high, palms out.

Use the left hand to grab the opponent's forearm and push the opponent's hand with the pistol away to clear the body from in front of the weapon. At the same time, rotate the right shoulder back to clear the body from the weapon.

Maintain control of the opponent's arm.

Grasp the weapon with the right hand by placing the thumb underneath the pistol and the fingers over top of the pistol.

Keep the right hand wrapped tightly around the muzzle and quickly rotate the pistol in the opponent's hand so the muzzle is facing the opponent.

Grasp and pull the opponent's wrist or forearm away from the body while rotating the weapon.

Rotate the weapon toward the opponent while pulling it up and back and out of the opponent's grasp.

Pistol to the Rear

This technique is performed when Marines are unarmed and the opponent has a pistol pointing to the back of the Marine's head. To execute the counter when the opponent is pointing a pistol to the rear, Marines—

Place the hands close together about chest high, palms out.

Step back with the left foot, pivoting on the right foot so the side is against the opponent's front. This action clears the body from the weapon's line of fire. Keep the left hand up.

Pivot on the left foot to face the opponent and, at the same time, raise the left elbow and reach over the top of the opponent's arm with the left arm.

Wrap the left arm tightly around the opponent's arm above his elbow to control it.

Push on the opponent's shoulder with the right hand while pulling up with the left arm to achieve an armbar. This action releases the opponent's grip on the weapon. If necessary, the opponent can be taken to the ground with a leg sweep.

Note: To execute this technique, the weapon must be close to or touching the rear of the Marine. If the weapon is too far away from the body, this technique would be difficult to execute or it would be ineffective.

CHAPTER 3

HAND-HELD WEAPONS

This chapter describes all techniques for a right-handed person. However, all techniques can be executed from either side.

In drawings, the Marine is depicted in woodland camouflage utilities; the opponent is depicted without camouflage. In photographs, the Marine is depicted in woodland camouflage utilities; the opponent is depicted in desert camouflage utilities.

Marines must know how to defend against attacks when an opponent is either unarmed or armed with a held-held weapon. This chapter addresses the combative use of knives, specific weapons of opportunity, and sticks. However, virtually anything can be used as a hand-held weapon.

1. Fundamentals of Knife Fighting

Marines must be trained in knife fighting techniques. Marines experienced in offensive knife techniques can cause enough damage and massive trauma to stop an opponent. When engaged against each other, experienced knife fighters employ various maneuvers and techniques that are specific to knife fighting. Seldom, if ever, will Marines engage an opponent in a classical knife fight.

Note: When armed with a rifle, Marines are issued a bayonet. When armed with a pistol, Marine are issued a combat knife.

Angles of Attack

There are six angles from which an attack with a knife can be launched:

- Vertical strike coming straight down on an opponent.
- Forward diagonal strike coming in at a 45-degree angle to the opponent.
- Reverse diagonal strike coming in at a 45-degree angle to the opponent.
- Forward horizontal strike coming in parallel to the ground.
- Reverse horizontal strike coming in parallel to the ground.
- Forward thrust coming in a straight line to the opponent.

Target Areas of the Body

During any confrontation, the parts of the opponent's body that are exposed or readily accessible will vary. The goal in a knife fight is to attack the body's soft, vital target areas that are readily accessible (e.g, the face, the sides and front of the neck, the lower abdomen [or groin]).

Neck. Carotid arteries, located on either side of the neck, are good target areas because they are not covered by body armor or natural protection.

Lower Abdomen (or Groin). The lower abdomen (or groin region) is a good target area because it is not covered by body armor.

Heart. The heart, if not covered by body armor, is an excellent target which, if struck, can prove fatal in a matter of seconds or minutes.

Secondary Targets. There are secondary target areas that will cause substantial bleeding if an artery is severed. These target areas are not immediately fatal, but can become fatal if left unattended. Attacks to—

The legs can cause a great deal of trauma and prove fatal. For example, the femoral artery located on the inside of the thigh is a large artery which, if cut, will cause extensive blood loss.
The brachial artery, located between the biceps and triceps on the inside of the arm, can cause extensive bleeding and damage.
The arm's radial and ulnar nerves can cause extensive bleeding and damage.

Movement

Marines can move anywhere within a 360-degree circle around the opponent. This allows accessibility to different target areas of the opponent's body. Marines should avoid being directly in front of an opponent because the opponent can rely on his forward momentum to seize the tactical advantage. If Marines face an opponent, movement is made in a 45-degree angle to either side of the opponent. This angle avoids an opponent's strike and places Marines in the best position to attack an opponent.

Wearing the Combat Knife

Marines must wear the combat knife where it is easily accessible and where it can best be retained. It is recommended the combat knife be worn on the weak side hip, blade down, sharp edge facing forward. Marines can place it behind the magazine pouch where it is easily accessible to them, but not easily grabbed by an opponent.

Grip

The grip on the knife should be natural. Marines grasp the knife's grip with the fingers wrapped around the grip naturally as it is pulled out of its sheath. This is commonly known as a hammer grip. The blade end of the knife is always facing the opponent.

Stance

Marines use the basic warrior stance as the foundation for knife techniques. The left hand is a vertical shield that protects either the ribs or the head and neck. The right elbow is bent with the blade pointing forward toward the opponent's head. This position serves as an index point, where all techniques are initiated.

Principles of Knife Fighting

The following are key principles of knife fighting:

- Execute movements with the knife blade within a box, shoulder-width across from the neck down to the waistline. The opponent has a greater chance of blocking an attack if the blade is brought in a wide, sweeping movement to the opponent.
- Close with the opponent, coming straight to the target.
- Move with the knife in straight lines.
- Point the knife's blade tip forward and toward the opponent.
- Apply full body weight and power in each of the knife techniques. Full body weight should be put into the attack in the direction of the blade's movement (slash or thrust).
- Apply constant forward pressure with the body and blade to keep the opponent off-balanced.

2. Knife Fighting Techniques

Slashing Techniques

Marines use slashing techniques to close with an enemy. Slashing techniques distract the opponent or damage the opponent so Marines can close in. Typically, Marines target the opponent's limbs, but any portion of the body that is presented can become a target.

Vertical Slash Technique. The vertical slash follows a vertical line straight down through the target. To execute the vertical slash, Marines—

Thrust the right hand out and bring the weapon straight down on the opponent, continuing to drag the knife down through the opponent's body.

Maintain contact on the opponent's body with the blade of the knife.

Forward Slash Technique. The forward slash follows a straight line in a forehand stroke, across the target areas of either the neck (high diagonal slash) or abdominal region (low horizontal slash). To execute the forward slash, Marines-

Extend the right hand while simultaneously rotating the palm up until the knife blade makes contact with the opponent.

Snap or rotate the wrist through the slashing motion to maximize blade's contact with the opponent.

Drag the knife across the opponent's body, from right to left, in a forehand stroke. The movement ends with the forearm against the body and the knife at the left hip with its blade oriented toward the opponent.

Reverse Slash Technique. The reverse slash is a follow-up technique to a forward attack. It allows Marines both a secondary attack and the ability to resume the basic warrior stance. The reverse slash follows a straight line in a backhand stroke, across the target areas of either the neck (high diagonal slash) or abdominal region (low horizontal slash). To execute the reverse slash, Marines—

Extend the right hand while simultaneously rotating the palm down until the knife blade makes contact with the opponent.

Snap or rotate the wrist through the slashing motion to maximize the blade's contact with the opponent.

Drag the knife across the opponent's body, from left to right, in a backhand stroke. Maintain contact on the opponent's body with the blade of the knife.

Thrusting Techniques

The primary objective of knife fighting is to insert the blade into an opponent to cause extensive damage and trauma. This is done with a thrusting technique. Thrusting techniques are more effective than slashing techniques because of the damage they can inflict. However, Marines use slashing techniques to close with the enemy so that they are closer to the opponent, which allows them to use the thrusting technique.

Vertical Thrust. The thrusting motion follows a vertical line straight up through the target (low into the abdomen region or high into the neck). To execute the vertical thrust, Marines—

Thrust the right hand toward the target, inserting the knife blade straight into the opponent.

Pull the knife out of the opponent.

Forward Thrust. The forward thrust follows a straight line straight into the opponent's neck (high thrust) or abdominal region (low thrust). To execute the forward thrust, Marines—

Thrust the right hand, palm down, toward the target, inserting the knife blade straight into the opponent.

Rotate the palm up once the knife is inserted to twist the blade.

Drop the right elbow and bring the knife to the opposite side of the opponent's body from where it was inserted. At the same time, rotate the hips and shoulders downward to bring body weight to bear on the attack.

Reverse Thrust. The reverse thrust is a follow-up technique to a forward attack. It allows Marines both a secondary attack and the ability to resume the basic warrior stance. The reverse thrust follows a horizontal line straight into the opponent's neck (high thrust) or abdominal region (low thrust). To execute the reverse thrust, Marines-

Bend the right arm, crossing the arm to the left side of the body.

Thrust the right hand, palm up, toward the target, inserting the knife blade straight into the opponent.

Rotate the palm down to twist the blade once the knife is inserted.

Bring the knife to the opposite side of the opponent's body from where it was inserted.

3. Weapons of Opportunity

During an unarmed close combat situation, Marines use their bodies as weapons, but they should be ready and able to use anything around them as a weapon. For example, Marines could throw sand or liquid in an opponent's eyes to temporarily impair his vision or smash the opponent's head with a rock or helmet. Marines must use whatever means are available and do whatever it takes to take control of the situation and to win, or they face the possibility of losing their lives. Some weapons of opportunity are discussed in the following subparagraphs.

Entrenching Tool

An entrenching tool (E-tool) is commonly carried by Marines. It can be an excellent weapon, especially when sharpened. Marines can use the E-tool to block, slash, and thrust at an opponent.

Tent Pole and Pins

Marines can use tent poles and pins to block, strike, or thrust at an opponent.

Web Belt

Marines can stretch a web belt between their hands to block attacks by an opponent.

Battlefield Debris

Marines can use debris on the battlefield (e.g., sticks, glass, a sharp piece of metal) to cut, slash, or stab an opponent. They can also use other types of debris such as shovel or ax handles, boards, metal pipes, or broken rifles to strike an opponent or apply a choke.

Helmet

A helmet can be used to strike an opponent on an unprotected area like the head and face. Grasp the rim of the helmet and thrust the arms forward, striking the opponent with the top of the helmet.

4. Fundamentals of Combative Stick

On the battlefield, Marines must be ready and able to use anything as a weapon. They must learn and be able to use techniques that can be employed with most weapons of opportunity. Among these techniques are combative stick techniques. Combative stick techniques can be used with a stick, a club, a broken rifle, an E-tool, or even a web belt.

Angles of Attack

There are six angles from which an attack with a hand-held weapon can be launched:

Vertical strike coming straight down on an opponent.
Forward diagonal strike coming in at a 45-degree angle to the opponent.
Reverse diagonal strike coming in at a 45-degree angle to the opponent.
Forward horizontal strike coming in parallel to the ground.
Reverse horizontal strike coming in parallel to the ground.
Forward thrust coming in a straight line to the opponent.

Grip

Grasp the stick about 2 inches from its base.

Stance

The basic warrior stance serves as the foundation for combative stick techniques. The left hand becomes a vertical shield that protects the ribs or the head and neck. Depending on how heavy the weapon is, it should be held at a level approximately shoulder height.

Movement

Movement during combative stick techniques is the same as it is for other close combat techniques. Marines can move anywhere within a 360-dcgrcc circle around the opponent. This allows accessibility to different target areas of the opponent's body and gains a tactical advantage. Marines should avoid being directly in front of an opponent because the opponent can rely on his forward momentum to seize the tactical advantage. If Marines face an opponent, movement is made in a 45-dcgrcc angle to cither side of the opponent. This angle avoids an opponent's strike and places Marines in the best position to attack an opponent.

5. Combative Stick Techniques

Strikes

Strikes are intended to inflict as much damage on an opponent as possible. Striking techniques apply to a weapon of opportunity such as a stick, a tent pole, a club, a broken rifle, an E-tool, or a Pipe-

Vertical Strike. To execute the vertical strike, Marines—

Bend the right arm, extending the weapon over the back of the right shoulder.

Rotate the forearm straight down off the elbow to bring the weapon down on the opponent.

Rotate the hips and shoulders forcefully toward the opponent.

Follow through with the strike by allowing the weight of the weapon to go through the target area of the body.

Forward Strike. To execute a forward strike, Marines—

Step forward with the left foot in the direction of the strike.

Bend the right arm with the elbow extending out to the right and the weapon extended over the right shoulder.

Rotate the forearm to the right of the elbow to bring the weapon down onto the opponent. At the same time, forcefully rotate the hips and shoulders toward the opponent.

Follow through with the strike by allowing the weight of the weapon to go through the target area of the body.

Reverse Strike. The reverse strike is a follow-up technique to a forward strike. It allows Marines both a secondary attack and the ability to resume the basic warrior stance.To execute a reverse strike, Marines—

Step forward with the right foot in the direction of the strike.

Bend the right arm with the hand near the left shoulder. The weapon is extended over the left shoulder.

Rotate the forearm to the right of the elbow to bring the weapon down onto the opponent. At the same time, forcefully rotate the hips and shoulders toward the opponent.

Follow through with the strike by allowing the weight of the weapon to go through the target area of the body.

Forward Thrust. To execute the forward thrust, Marines—

Grasp the stick with the left hand, palm up, in a position where the stick can be controlled with two hands.

Lift the left leg and lunge forward off the ball of the right foot. At the same time, thrust the end of the weapon directly toward the opponent by thrusting both hands forward in a straight line.

6. Blocking Techniques

A block is meant to deter or deflect an attack by an opponent. A block sets up Marines for a follow-on attack against the opponent. Blocks are executed by deflecting, rather than hitting or following through like a strike.

Blocks Against Unarmed Attacks

To block against an unarmed attack, Marines—

Step forward at a 45-degree angle with either the right or left foot. This moves the body out of the line of attack.

Raise the left arm and block the strike with the meaty portion of the forearm.

Employ the stick by one of two techniques:

Strike the opponent with the stick.
Use the stick to block on two points of contact. When the stick is used to block, it serves as an incidental strike. This technique is only used if Marines have closed with the opponent and are inside his strike.

Blocks Against Armed Attacks

Block for a Vertical Strike. To execute the block for a vertical strike, Marines—

Step forward with the left foot at a 45-degree angle to the left. This moves the body out of the line of attack.

Block on two points of contact to disperse the impact of the attack:

Block the opponent's stick by positioning the stick so it is perpendicular to the opponent's stick. If the stick is not perpendicular to the opponent's stick, the stick can slide through and make contact on the Marine.
Block the opponent's wrist or forearm with the back of the left forearm.

Note: Use the stick to block the opponent's arm if closer to the opponent. It is the same movement as blocking with the arm, except the opponent's arm is blocked with both the stick and the arm.

Block for a Forward Strike. To execute the block for a forward strike, Marines-
Step forward with the left foot at a 45-degree angle to the left. This moves the body out of the line of attack and inside the opponent's strike.

Block on two points of contact to disperse the impact of the attack:

- Block the opponent's wrist or forearm with the meaty portion of the back of the left forearm.
- Strike the opponent's attacking biceps with the stick.

Block for a Reverse Strike. To execute the block for a reverse strike, Marines—

Step forward at a 45-degree angle with the right foot to the right. This moves the body out of the line of attack and inside the opponent's strike.

Block on two points of contact to disperse the impact of the attack:

Block the opponent's stick by positioning the stick so it is perpendicular to the opponent's stick.

Block the opponent's forearm with the meaty portion of the left forearm.

Note: If closer to the opponent, block his triceps with the back of the left forearm and strike his forearm with the stick.

7. Unarmed Against Hand-Held Weapons

If Marines are engaged against an opponent with a knife, a stick, or some other weapon of opportunity, they must establish and maintain an offensive mindset, not a defensive mindset. Their survival depends on it. They cannot afford to think about getting cut or hurt.

Angles of Attack

Before Marines can learn to block or counter an attack with a hand-held weapon (e.g., knife, stick), they must know from what angle the opponent is attacking. There are six angles from which an opponent will typically attack with a hand-held weapon:

Vertical strike coming in straight down on the Marine.
Forward diagonal strike coming in at a 45-degree angle to the Marine.
Reverse diagonal strike coming in at a 45-degree angle to the Marine.
Forward horizontal strike coming in parallel to the ground.
Reverse horizontal strike coming in parallel to the ground.
Forward thrust coming in a straight line to the Marine.

Blocks

If the opponent has a hand-held weapon (e.g., knife or stick), Marines parry the opponent's hand or arm to block the attack.

Basic Block. To execute the basic block technique, Marines-

Step forward at a 45-degree angle to move out of the line of the attack. Always step in the direction of the strike.

Thrust the forearms forward, hands up, against the opponent's attacking arm. Contact is made on the opponent's arm with the backs of the fore-arms.

Block for a Vertical Strike. To execute the block against a vertical strike, Marines—

Step forward with the left foot at a 45-degree angle to the left to move out of the line of attack.

Thrust the forearms forward, hands up, against the outside of the opponent's attacking arm.

Block for a Forward Strike. To execute the block against a forward diagonal or forward horizontal strike, Marines-

Step forward with the left fool inside the opponent's attacking arm.

Block the attack with both arms bent so the fore-arms make contact with the opponent's biceps and forearm.

Block for a Reverse Strike. To execute the block against a reverse diagonal or reverse horizontal strike, Marines—

Step forward with the right foot to the outside of the opponent's attacking arm.

Block the attack with both arms bent so the fore-arms make contact with the opponent's triceps and forearm.

Block for a Forward Thrust. To execute the block against a forward thrust, Marines—

Bend at the waist, move the hips backward, and jump backward with both feet to move away from the attack. This action is known as "hollowing out."

Hollow out and block the attack with the arms bent and hands together on top of the attacking arm.

Overlap the hands slightly so one thumb is on top of the other hand's index finger. The other thumb should be under the other hand's index finger.

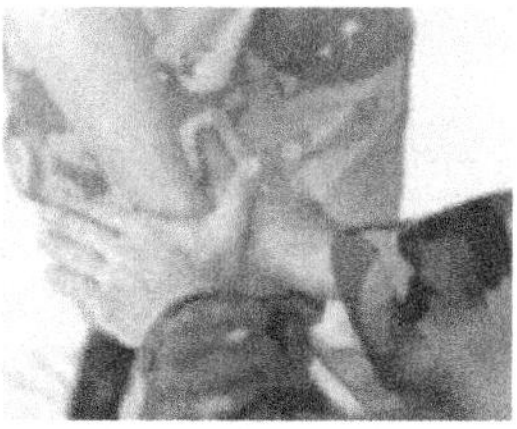

8. Counters to Hand-Held Weapon Attacks

Principles of Counters

A counter is used to control the situation to regain the tactical advantage and end the fight. Regardless of the type of weapon or angle of attack, the following principles apply to countering the attack with a hand-held weapon:

Move out of the line of attack. Movement is executed in a 45-degree angle forward to the left or right.

- Block the attack.

Note: The first two actions are taken simultaneously.

- Take control of the weapon by controlling the hand or arm that is holding the weapon. Never attempt to grab the opponent's weapon.
- Execute the appropriate follow-up to end the fight; e.g., strikes, joint manipulations, throws, or takedowns (see chap. 8). Marines should continue to attack the opponent until the fight ends.

Counter Techniques

There are two techniques that can be used to counter any armed attack: forward armbar counter and reverse armbar counter. These techniques can be used to counter a vertical attack, a forward diagonal strike, or a forward horizontal strike. With minor variations, the same techniques arc used to counter reverse strikes. A third technique, the bent armbar counter, is used to counter a vertical attack.

Forward Armbar Counter. To execute the forward armbar counter to an attack coming from a forward strike, Marines-

Step forward with the left foot inside the opponent' s attacking arm.

Block the attack with both arms bent so the forearms make contact with the opponent's biceps and forearm.

Slide the left arm over the opponent's forearm and wrap the arm tightly around his arm, trapping the opponent's attacking arm between the biceps and torso.

Place the right hand on the opponent's shoulder or upper arm to further control his arm and to effect an armbar.

Execute an armbar and continue to exert steady pressure against the arm.

Reverse Armbar Counter. To execute the reverse armbar counter to an attack coming from a forward strike, Marines-

Step forward with the left foot inside the opponent' s attacking arm.

Block the attack with both arms bent so the forearms make contact with the opponent's biceps and forearm.

Control the opponent's arm with the left hand and pivot to the right so the back is against the opponent's side. Immediately slide the right arm over the opponent's biceps and wrap the arm tightly around his arm, trapping the opponent's attacking arm between the biceps and torso.

Grasp the opponent's wrist with the left hand and twist his thumb away from the body.

Take the opponent to the ground with an armbar takedown if he does not drop the weapon from the wristlock.

Bent Armbar Counter. This counter is particularly effective against a vertical attack. To execute the bent armbar counter, Marines—

Step forward with the left foot inside the opponent's attacking arm.

Block the attack with both arms bent so the forearms make contact with the opponent's biceps and forearm.

Grasp the opponent's forearm with the left hand. At the same time, slide the right arm underneath the opponent's triceps and grasp the opponent's forearm or wrist with the right hand.

Apply downward pressure with the hands against the opponent's forearm to off-balance the opponent.

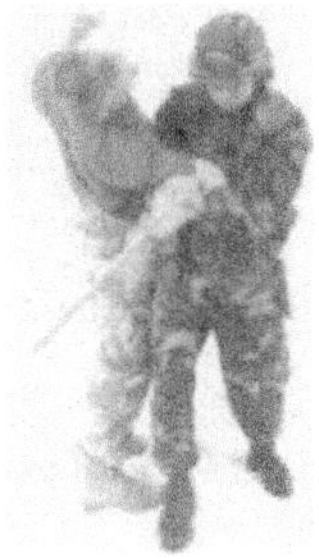

(reverse blank)

CHAPTER 4

STRIKES

This chapter describes all techniques for a right-lmnded person. However, all techniques can be executed from either side.

In drawings, the Marine is depicted in woodland camouflage utilities; the opponent is depicted without camouflage. In photographs, the Marine is depicted in woodland camouflage utilities; the opponent is depicted in desert camouflage utilities.

Strikes are unarmed individual hitting techniques. Strikes use the hands, elbows, knees, feet, and, in some instances, other parts of the body as personal weapons. Marines must know how to execute strikes effectively. They must also know how to counter strikes from an opponent.

1. Principles of Punches

Muscle Relaxation

Muscle relaxation is crucial when executing punches. The natural tendency in a fight is to tense up, which results in rapid fatigue and decreased power generation. Marines who remain relaxed during a close combat situation generate greater speed, which results in greater generation of power. Relaxing the forearms generates speed and improves reaction time. At the point of impact, Marines clench the fist to cause damage to the opponent and avoid injury to the wrist and hand.

Weight Transfer

Weight transfer is necessary to generate power in a punch. Marines accomplish this by-

Rotating their hips and shoulders into the attack.

Moving their body mass straight forward or backward in a straight line.

Dropping their body weight into an opponent. Body mass can be transferred into an attack from high to low or from low to high.

Rapid Retraction

When Marines deliver a punch, rapid retraction of the fist is important. Once the hand has made contact with the target, Marines quickly return to the basic warrior stance. Rapid retraction-

Returns the hand and arm to the protection afforded by the basic warrior stance.

Prevents the opponent from being able to grab the hand or arm.

Permits the hand and arm to be "chambered" or "re-cocked" in preparation for delivering a subsequent punch.

Telegraphing

Telegraphing a strike occurs when body movements inform the opponent of the intention to launch an attack. Staying relaxed helps to reduce telegraphing.

Often, an untrained fighter telegraphs his intention to attack by drawing his hand back in view of his opponent, changing facial expression, tensing neck muscles, or twitching. These movements, however small, immediately indicate an attack is about to be delivered. If the opponent is a trained fighter, he may be able to evade or counter the attack. If the opponent is an untrained fighter, he may still be able to minimize the effect of an attack.

2. Punches

Punches may be thrown during any hand-to-hand confrontation. Most people resort to punching because it is a natural reaction to a threat. The purpose of a punch is to stun the opponent or to set him up for a follow-up finishing technique. However, punches should only be executed to the soft tissue areas of an opponent. A correctly delivered punch maximizes the damage to an opponent while minimizing the risk of injury to Marines.

Basic Fist

Punches are executed using the basic fist. To make the basic fist, the fingers are curled naturally into the palm of the hand and the thumb is placed across the index and middle fingers. Do not clench the fist until movement has begun. This reduces muscular tension in the forearm and increases speed and reaction time. Just before impact, Marines exert muscular tension on the hand and forearm to maximize damage to the opponent and reduce their chances of injury. Contact should be made with the knuckles of the index and middle fingers.

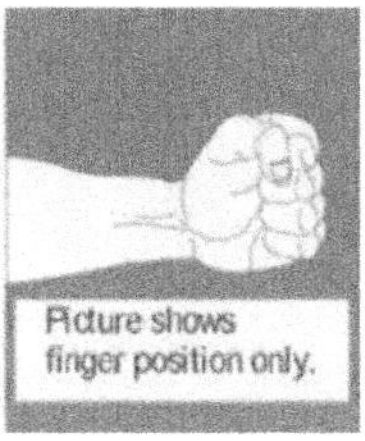

When striking with the basic fist, Marines must keep the hand straight, or in line, with the wrist to avoid injury to the wrist.

Lead Hand Punch

The lead hand punch is a snapping straight punch executed by the forward or lead hand. It is a fast punch designed to keep the opponent away and to set up an attack. A lead hand punch conceals movement and allows Marines to get close to the opponent. Lead hand punches should strike soft tissue areas, if possible. To execute the lead hand punch, Marines-

Snap the lead hand out to nearly full extension, while rotating the palm to the ground.

Contact the opponent with the first two knuckles of the fist.

Retract the hand immediately, resuming the basic warrior stance.

Rear Hand Punch

The rear hand punch is a snapping punch executed by the rear (right) hand. It is a power punch designed to inflict maximum damage on the opponent. Its power comes from pushing off the rear leg and rotating the hips and shoulders. To execute the rear hand punch, Marines—

Rotate the hips and shoulders forcefully toward the opponent and thrust the rear hand straight out, rotating the palm down, to nearly full extension.

Shift body weight to the lead foot while pushing off on the ball of the rear foot.

Contact the opponent with the first two knuckles of the fist.

Retract the hand immediately.

Uppercut

The uppercut is a powerful punch originating below the opponent's line of vision. It is executed in an upward motion traveling up the centerline of the opponent's body. It is delivered in close and usually follows a preparatory strike that leaves the target area unprotected. When delivered to the chin or jaw, the uppercut can render an opponent unconscious, cause extensive damage to the neck, or sever the tongue. To execute the uppercut, Marines—

Bend the arms, rotating the palm inboard. The distance the arms bend depends on how close the opponent is.

Rotate the hips and shoulders forcefully toward the opponent, thrusting the fist straight up toward the opponent's chin or jaw.

Contact the opponent with the first two knuckles of the fist.

Retract the hand immediately.

Hook

The hook is a powerful punch that is executed close in and is usually preceded by a preparatory strike. To execute the hook, Marines—

Thrust the right arm in a hooking motion toward the opponent, keeping the elbow bent while forcefully rotating the right shoulder and hip toward the opponent.

Contact the opponent with the first two knuckles of the fist. Continue rotating the shoulder and hip, following through with the fist to the target.

Retract the hand immediately.

3. Strikes with the Upper Body

Strikes stun the opponent or set him up for a follow-up finishing technique. The hands, forearms, and elbows are individual weapons of the arms that can be used to execute strikes including the hammer fist, chin jab, knife hand, eye gouge, and elbow strikes. These strikes provide a variety of techniques that can be used in any type of close combat situation.

Principles of Execution

There are several principles of execution that ensure a strike's effectiveness.

Generating Power. Weight transfer is necessary to generate maximum power in a punch. Marines accomplish this by-

- Rotating the hips and shoulders into the attack.
- Moving the body mass straight forward or backward in a straight line.
- Dropping body weight into an opponent. Body mass can be transferred into an attack from high to low or from low to high.

Muscular Tension. The arms are relaxed until the moment of impact. At the point of impact, Marines apply muscular tension in the hand and forearm to maximize damage to the opponent and to avoid injury to the hand. The arms are relaxed until the moment of impact.

Hit and Stick and Follow-Through. A strike should be delivered so that the weapon (e.g., hand, elbow) hits and remains on the impact site (target) and follows through the target. This technique inflicts maximum damage on the opponent.

Strikes with the arms are executed with "heavy hands"; i.e., the strike is executed by driving through with the strike to allow the weight of the hand to go through the target area of the body. Contact with the opponent should be made with the arm slightly bent, and the arm extends as it moves through the target. This technique allows Marines to deliver strikes effectively without executing full force.

Movement. Movement puts Marines in the proper position for launching an attack against an opponent as well as providing protection.

Movement is initiated from the basic warrior stance and ends by resuming the basic warrior stance. Strikes can be performed with either the left or right arm depending upon—

The angle of attack.
The position of the opponent.
The opponent's available, vulnerable target areas.

Target Areas of the Body

For each strike, there are target areas of the body that, when struck, maximize damage to an opponent. Strikes use gross motor skills as opposed to fine motor skills. The target areas of the body are just that—areas. Pinpoint accuracy on a specific nerve is not needed for the strike to be effective.

Hammer Fist

Striking with the hammer fist concentrates power in a small part of the hand, which, when transferred to the target, can have a devastating effect. The striking surface of the hammer fist is the meaty portion of the hand below the little finger. To execute the hammer fist strike, Marines-

Make a fist and bend the arm at approximately a 45- to 90-degree angle. At the same time, rotate the right hip and right shoulder backward.

Thrust the fist forward onto the opponent while rotating the right hip and shoulder forward.

Rotate the wrist so the hammer fist makes contact on the opponent.

Chin Jab

The chin jab can immediately render an opponent unconscious and cause extensive damage to the neck and spine. The striking surface is the heel of the palm of the hand. To execute the chin jab, Marines—

Bend the right wrist back at a 90-degree angle with the palm facing the opponent and the fingers pointing up.

Keep the right arm bent and close to the body. Extend the hand into a concave position with the fingers slightly spread apart.

Step forward with the left foot toward the opponent, keeping the feet approximately shoulder-width apart and the knees bent. This is done to close with the opponent.

Keep the right arm bent and close to the side. Thrust the palm of the hand directly up under the opponent's chin. At the same time, rotate the right hip forward to drive the body weight into the attack to increase the power of the strike. The attack should travel up the centerline of the opponent's chest to his chin.

Knife Hand

The knife hand is one of the most versatile and devastating strikes. The striking surface is the cutting edge of the hand, which is the meaty portion of the hand below the little finger extending to the top of the wrist. The striking surface is narrow, allowing strikes on the neck between the opponent's body armor and helmet. The knife hand strike is executed from one of three angles: outside, inside, and vertical.

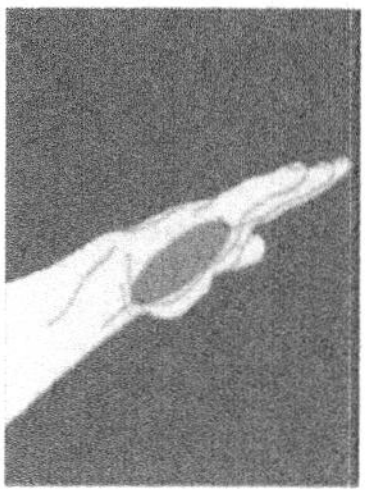

Outside Knife Hand. To execute the outside knife hand strike, Marines—

Execute a knife hand by extending and joining the fingers of the right hand and placing the thumb next to the forefinger (like saluting).

Retract the right hand. At the same time, rotate the right hip and right shoulder backward.

Thrust the knife hand forward (horizontally) onto the opponent while rotating the right hip and shoulder forward.

Inside Knife Hand. To execute the inside knife hand strike, Marines—

Execute a knife hand.

Bring the right hand over the left shoulder. At the same time, rotate the right shoulder forward and the left hip forward.

Thrust the knife hand forward (horizontally) onto the opponent while rotating the right hip and shoulder forward and the left shoulder backward.

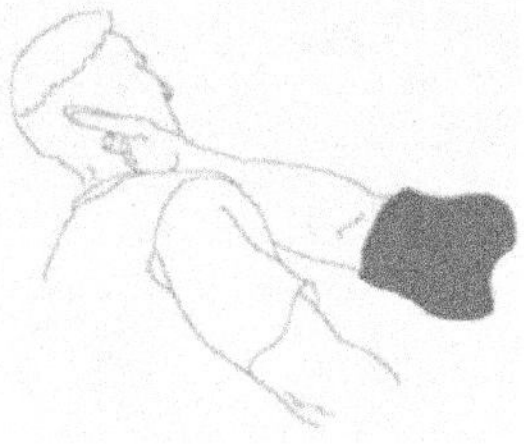

Vertical Knife Hand. When thrown vertically, the knife hand strike comes straight down in a straight line.

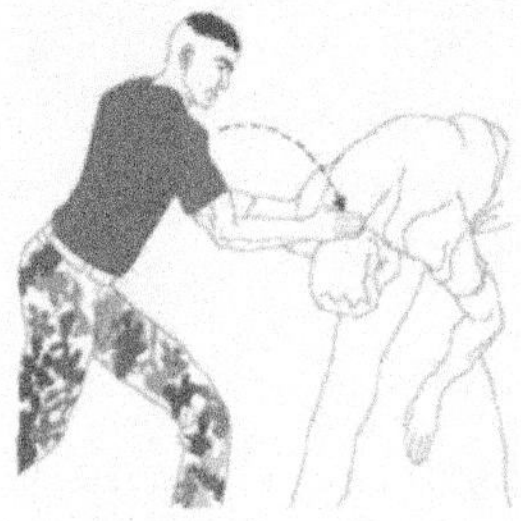

Eye Gouge

The eye gouge is used to attack an opponent's eyes, blinding him so follow-up strikes can be executed. The striking surface is the tips of the fingers or thumb. To execute the eye gouge, Marines—

Extend the right hand with the fingers slightly spread apart to allow entry into the eye sockets.

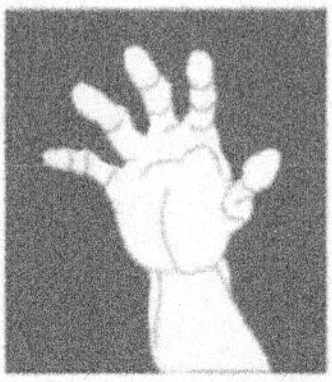

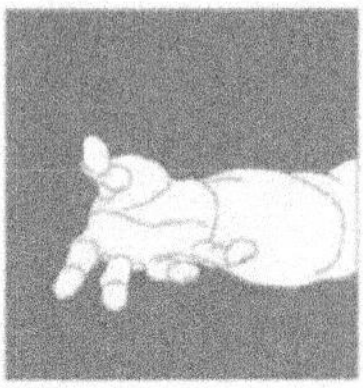

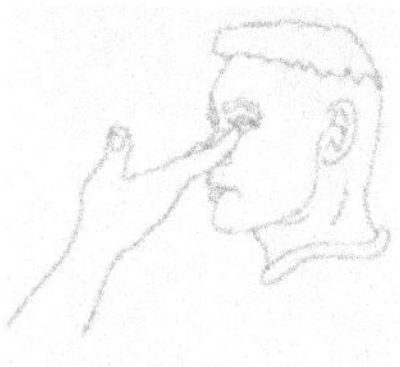

Place the palm of the hand either toward the ground or toward the sky and thrust the right hand forward into the opponent's eyes.

Thrust the hand forward at the opponent's nose level so the fingers or thumb slide naturally into the grooves of the opponent's eye sockets.

Elbow Strikes

The elbow is a powerful weapon that can be used in several different ways to attack virtually any part of an opponent's body. Elbow strikes can be performed either vertically (upward or downward) or horizontally (forward or reverse). The striking surface is 2 inches above or below the point of the elbow, depending upon the angle of attack, the opponent's attack angle, and the position of the opponent.

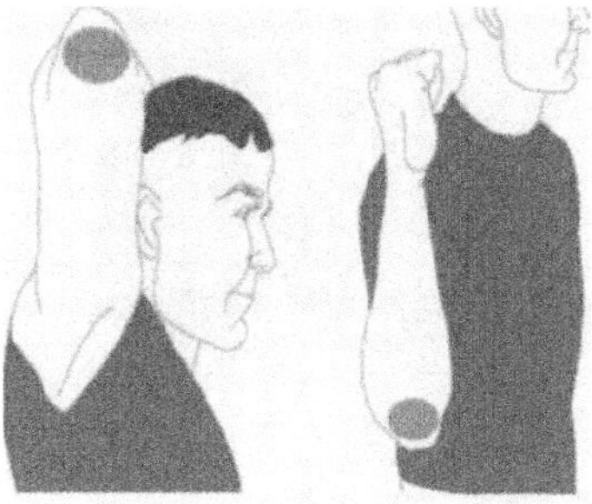

Above Elbow **Below Elbow**

Vertical Elbow Strike (Up). To execute an upward vertical elbow strike, Marines-

Bend the right elbow, keeping the fist close to the body. The fist is at shoulder level and the elbow is next to the torso.

Thrust the elbow vertically upward toward the opponent while rotating the right shoulder and hip forward to generate additional power.

Contact the opponent with the right forearm 2 inches above the point of the elbow.

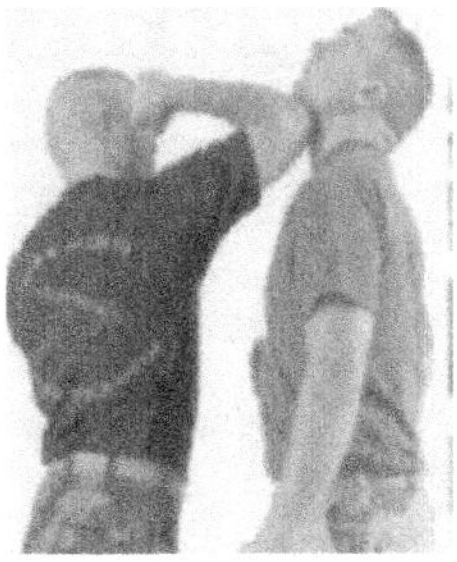

Vertical Elbow Strike (Down). To execute a downward vertical elbow strike, Marines—

Bend the right elbow, keeping the fist close to the body. The fist is on the shoulder and the elbow is raised well above the shoulder.

Thrust the elbow vertically downward toward the opponent while dropping body weight into the attack to generate additional power.

Contact the opponent with the right triceps 2 inches above the point of the elbow.

Horizontal Elbow Strike (Forward). To execute the forward horizontal elbow strike, Marines-

Tuck the right fist near the chest with the palm heel facing the ground.

Thrust the right elbow horizontally forward toward the opponent. The forearm is parallel to the ground.

Rotate the right shoulder and hip forward to generate additional power.

Contact the opponent with the right forearm 2 inches below the point of the elbow.

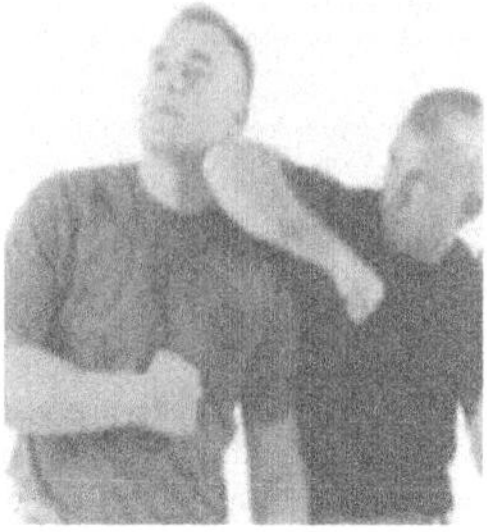

Horizontal Elbow Strike (Rear). To execute the rear horizontal elbow strike, Marines-

Tuck the right fist near the left shoulder with the palm heel facing the ground. At the same time, rotate the right shoulder forward and the left hip forward.

Thrust the right elbow horizontally rearward toward the opponent. The forearm is parallel to the ground and the hand moves toward the direction of the attack.

Rotate the right hip back and the right shoulder backward to generate additional power.

Contact the opponent with the right triceps 2 inches above the point of the elbow.

4. Strikes with the Lower Body

The legs are the body's most powerful weapons because they use the largest muscle groups to generate a strike. Legs are also less prone to injury. The feet are the preferred choice for striking because they are protected by boots. Marines use their feet, heels, and knees to execute kicks, knee strikes, and stomps.

Kicks

The purpose of a kick is to stop an opponent's attack or to create an opening in his defense in order to launch an attack. Kicks can be performed with the left (lead) leg or the right (rear) leg. Kicks with the rear leg have greater power because the hips are rotated into the attack. However, the rear leg is further away from the opponent so a strike with the rear leg will not contact the opponent as quickly as a strike with the lead leg.

Front Kick. The front kick is executed when the opponent is in front of the Marine. The front kick, delivered with either the rear or lead leg, is effective for striking below the waist. Attempting to kick higher results in diminished balance and provides the enemy with a greater opportunity to grab the leg or foot. The striking surfaces are the toe of the boot or the bootlaces. To execute the front kick, Marines-

Raise the left knee waist high, pivot the hips into the attack, and thrust the left foot forward toward the opponent.

Contact the opponent with the toe of the left boot or bootlaces.

Return to the basic warrior stance.

Side Kick. The side kick, delivered with the lead leg, is effective for striking the knees. The side kick is executed when the opponent is to the side of the Marine. The striking surface is the outside cutting edge of the boot near the heel. To execute the side kick, Marines—

Raise the right knee waist high and rotate the right hip forward.

Thrust the right foot to the right side toward the opponent, turning the foot at a 90-degree angle to maximize the striking surface on the opponent.

Contact the opponent with the cutting edge of the right boot.

Return to the basic warrior stance.

Knee Strikes

Knee strikes are excellent weapons during the close range of close combat fighting. The knee strike is generally delivered in close.

Vertical Knee Strike. The striking surface is the thigh, slightly above the knee. To execute the vertical knee strike, Marines-

Raise the right knee and drive it up forcefully into the opponent. Power is generated by thrusting the leg upward.

Contact the opponent 2 inches above the right knee.

Horizontal Knee Strike. The horizontal knee strike is executed with the leg generally parallel with the ground while rotating the hips to generate power. It is often delivered to the peroneal nerve. The striking surface is the front of the leg, slightly above or below the knee. To execute the horizontal knee strike, Marines-

Raise the right knee, rotate the right hip forward while pivoting on the left foot, and drive the knee horizontally into the opponent.

Contact the opponent 2 inches above the right knee.

Stomps

Stomps are delivered with the feet, usually when the opponent is down. Remember, when the opponent is down, Marines take whatever target is available.

Vertical Stomp. The vertical stomp allows Marines to remain upright and balanced, to rapidly deliver multiple blows with either foot, and to quickly and accurately attack the target. It is the preferred stomp. The striking surface is the flat bottom of the boot or the cutting edge of the heel. To execute the vertical stomp, Marines—

Raise the right knee above the waist with the right leg bent at approximately a 90-degree angle.

Drive the flat bottom of the right boot or the cutting edge of the right heel down onto the opponent forcefully. At the same time, bend the left knee slightly to drop the body weight into the strike.

Ax Stomp. The striking surface of the ax stomp is the cutting edge of the heel. To execute the ax stomp, Marines-

Raise the right heel above the waist, keeping the right leg straight.

Drive the cutting edge of the right heel down onto the opponent forcefully. At the same time, bend the left knee slightly to drop the body weight into the strike.

5. Counters to Strikes

In a close combat situation, an opponent will attempt to strike Marines with punches and kicks. When an opponent uses a strike, Marines must first avoid the strike. This is accomplished by moving quickly and blocking. Next, Marines must get into an offensive position. This allows Marines to use offensive strikes to attack the opponent. Regardless of the strike, the counter to a strike requires Marines to move, block, and strike.

Move

The first step in countering a strike is to move out of the way of the strike's impact. Movement removes Marines from the opponent's intended strike point and positions Marines to attack. Movement is executed at approximately a 45-degree angle to the front or rear. Movement is always initiated from the basic warrior stance. Return to the basic warrior stance with the toe of the lead foot pointing toward the opponent once the movement is complete.

Block

Different blocks are executed based on the strike. These will be covered by individual counters.

Strike

Any of the upper body or lower body strikes can be executed as a follow-on attack or part of the counter to an opponent's strike. The follow-on strike is determined by the angle to the opponent, the position of the opponent, and the opponent's available vulnerable target areas.

Counters to Punches

Counter to a Lead Hand Punch. This counter is used when the opponent throws a lead hand punch. To execute the counter to the lead hand punch, Marines-

Step forward and to the left at approximately a 45-degree angle, moving in to the outside of the opponent's attacking arm.

Raise the left arm and block or deflect the opponent's lead hand with the palm of the hand or the meaty portion of the forearm.

Hit and stick by leaving the left arm against the opponent's right arm while stepping forward and to the left at approximately a 45-degree angle to close with the opponent.

Execute a finishing technique, such as a strike or a kick, to the opponent's exposed target areas.

Counter to a Rear Hand Punch. This counter is used when the opponent throws a rear hand punch. To execute the counter to the rear hand punch, Marines—

Step forward and to the left at approximately a 45-degree angle, moving in to the outside of the opponent's attacking arm.

Raise the left arm and block or deflect the opponent' s rear hand with the palm of the hand or the meaty portion of the forearm.

Hit and stick by leaving the left arm against the opponent's right arm while stepping forward and to the right at approximately a 45-degree angle to close with the opponent.

Execute a finishing technique, such as a strike or a kick, to the opponent's exposed target areas.

Counters to Kicks

Counter to a Front Kick (Left or Lead Leg). This counter is used when the opponent executes a front kick with his left leg. To execute the counter to a front kick, Marines-

Step forward and to the right at approximately a 45-degree angle, moving in to the outside of the opponent's striking leg.

Lower the left arm and block or deflect the opponent's leg with the palm of the hand or the meaty portion of the forearm.

Hit and stick by leaving the left arm against the opponent's leg while stepping forward and to the left at approximately a 45-degree angle to close with the opponent.

Execute a finishing technique, such as a strike or a kick, to the opponent's exposed target areas.

Counter to a Front Kick (Right or Rear Leg). This counter is used when the opponent executes a front kick with his right leg. To execute the counter to a front kick, Marines—

Step forward and to the left at approximately a 45-degree angle, moving in to the outside of the opponent's striking leg.

Lower the left arm and block or deflect the opponent's leg with the palm of the hand or the meaty portion of the forearm.

Hit and stick by leaving the left arm against the opponent's leg while stepping forward and to the right at approximately a 45-degree angle to close with the opponent.

CHAPTER 5

THROWS

This chapter describes all techniques for a right-handed person. However, all techniques can be executed from either side.

The Marine is depicted in camouflage utilities. The opponent is depicted without camouflage.

Marines use throwing techniques to maintain the tactical advantage and to throw the opponent to the ground during close combat. Throws apply the principles of balance, leverage, timing, and body position to upset an opponent's balance and to gain control by forcing the opponent to the ground. Throwing techniques are effective because they are size- and gender-neutral, and they rely on the momentum and power generated by the opponent rather than the strength or size of the Marine. Marines also execute a throw as a devastating attack against an opponent, possibly causing unconsciousness or broken limbs. When Marines execute throws, they must maintain balance and, simultaneously, prevent the opponent from countering a throw or escaping after being forced to the ground.

1. Turning Throw

Marines use a turning throw to take the opponent to the ground while they remain standing. A turning throw can also be executed from a stationary position. It is particularly effective if Marines and the opponent are wearing gear. To execute the turning throw, Marines-

Grasp the opponent's right wrist with the left hand.

Step forward with the right foot, place it against the outside of the opponent's right foot, and pivot so the back of the heel is next to the middle of the opponent's foot.

Hook the opponent's right arm with the right arm and pinch his arm between the biceps and forearm, touching the opponent with the body.

Pull the opponent's wrist downward, keeping it close to the body.

Pivot to the left on the ball of the foot and continue pulling downward on the opponent's wrist while rotating the wrist outward to off-balance him.

2. Hip Throw

Marines use a hip throw to take the opponent to the ground while they remain standing. A hip throw is particularly effective if the opponent is moving forward or pushing. Marines use the opponent' s forward momentum to execute the hip throw. To execute the hip throw, Marines-Grasp the opponent's right wrist with the left hand.

Step forward with the right foot and place it against the outside of the opponent's right foot.

The back of the heel should be next to the opponent' s foot.

Step back with the left foot and rotate on the ball of the foot. The back of the heel is next to the opponent' s toe. The knees are bent.

Rotate at the waist and hook the right hand around the back of the opponent's body (anywhere from his waist to his head). The side and hip should be against the opponent.

Rotate the hip against the opponent. The hips must be lower than the opponent's.

Use the right hand to pull the opponent up on the hip to maximize contact.

Pull the opponent's arm across the body and, at the same time, lift the opponent off the ground slightly by bending at the waist, straightening the legs, and rotating the body to the left.

3. Leg Sweep

Marines use a leg sweep to take the opponent to the ground while they remain standing. A leg sweep is particularly effective if the opponent is already off-balanced and moving backward or pulling on the Marine. To execute the leg sweep, Marines—

Grab the opponent's right wrist with the left hand and grab the opponent' s left shoulder with the right hand.

Note: Grab the opponent's clothing or gear if his wrist and shoulder cannot be grabbed.

Step forward with the left foot and place it on the outside of the opponent's right foot.

Pull the opponent's wrist downward, close to the body, and push his shoulder backward to off-balance him.

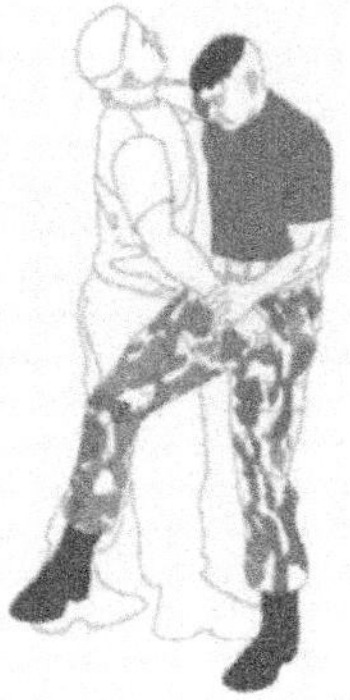

Raise the right knee no higher than the waist.

Kick the foot past the opponent's right leg.

Use the heel of the boot to make contact with the opponent's calf (anywhere from the top of the calf down to the Achilles tendon or on the inside of the calf).

Sweep the opponent to the ground.

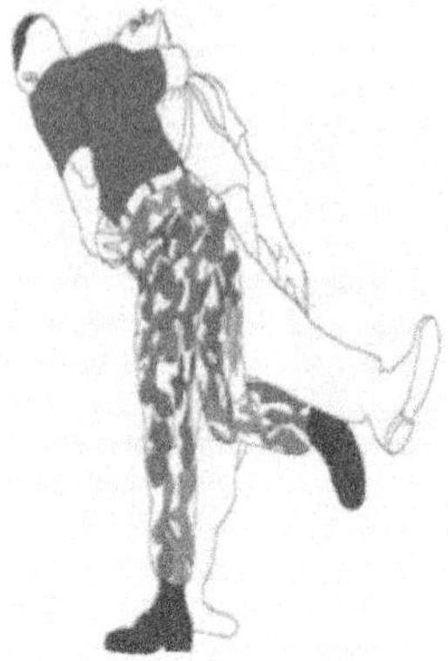

CHAPTER 6

CHOKES AND HOLDS

This chapter describes all techniques for a right-handed person. However, all techniques can be executed from either side.

The Marine is depicted in camouflage utilities. The opponent is depicted without camouflage.

WARNING

During training, never execute a choke at full force or full speed and never hold a choke for more than 5 seconds.

When Marines correctly perform a choke, they render an opponent unconscious in as little as 8 to 13 seconds. Chokes are easily performed regardless of size or gender. Marines must know how to apply chokes and how to counter a choke or a hold executed by an opponent.

1. Types of Chokes

There are two types of chokes: an air choke and a blood choke. An air choke closes off the airway to the lungs, thereby preventing oxygen from reaching the heart. A blood choke cuts off the blood flow to the brain. Both types can result in unconsciousness and eventual death for an opponent.

Air Choke

An air choke is performed on the opponent's windpipe (or trachea), cutting off the air to the lungs and heart. If Marines execute the air choke properly, the opponent loses consciousness within 2 to 3 minutes. Due to the length of time it takes to immobilize the opponent, air chokes are not recommended.

Blood Choke

A blood choke is performed on the opponent's carotid artery, which carries oxygen-enriched blood from the heart to the brain. The carotid artery is located on both sides of the neck. If Marines execute a blood choke properly, the opponent will lose consciousness within 8 to 13 seconds. The blood choke is the preferred choke because its intended effect (i.e., the opponent losing consciousness) can be executed quickly, ending the fight.

2. Chokes

Front Choke

Marines execute a front choke when they are facing the opponent. Marines use the opponent's lapels or collar to execute a front choke. To execute the front choke, Marines-

Grab the opponent's right lapel with the right hand, making certain that the knuckles or the back of the hand are against the opponent's carotid artery on the right side of his neck.

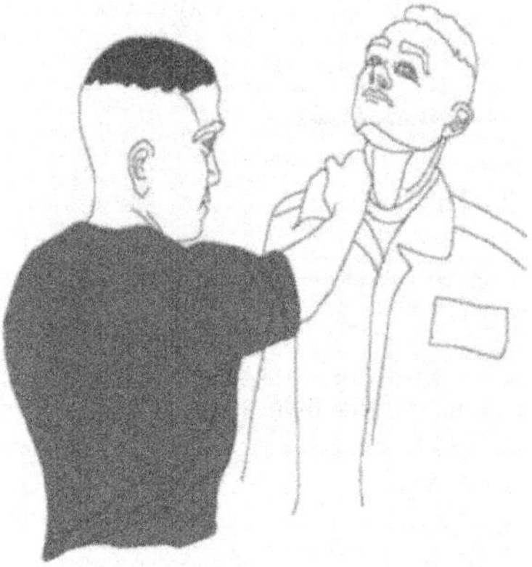

Keep the right hand against the opponent's neck, reach under the opponent's right arm with the left hand, grab the opponent's left lapel, and form an X with the wrists.

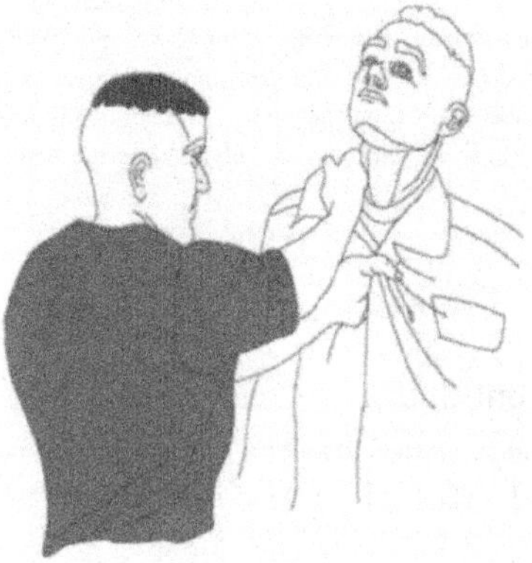

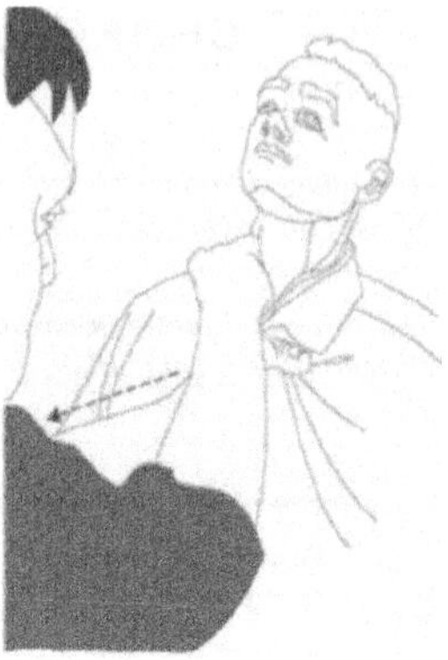

Press the right hand against the opponent's carotid artery.

Pull his lapel to the left with the left hand.

Side Choke

Marines execute a side choke when they are facing the opponent. The side choke is particularly effective when deflecting a punch thrown by an opponent. To execute the side choke, Marines—

Use the left hand to parry the opponent's arm inboard (to the inside of the opponent's reach).

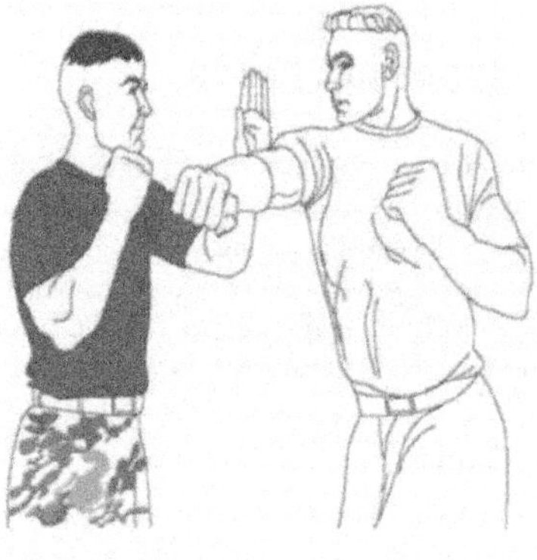

Bring the right arm underneath the opponent's arm and up around the front of his neck.

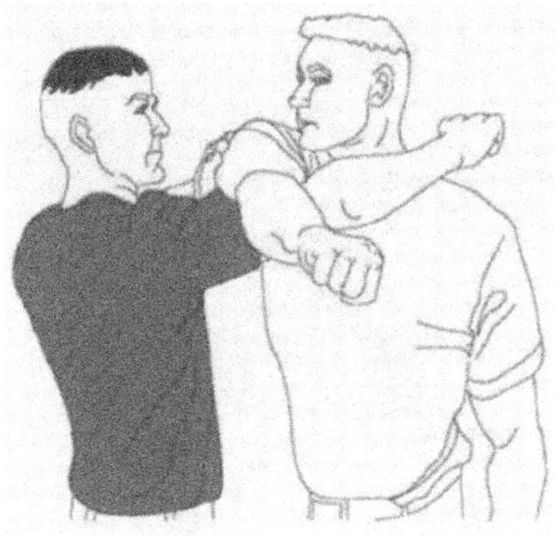

Extend the fingers, place the back of the forearm against the opponent's neck just below his ear, and press the carotid artery on his neck.

Note: If Marines are unable to place their fingers against the opponent's neck, they may use the back of the thumb or wrist.

Reach, with the left hand, around the back of the opponent's neck and clasp the hands together.

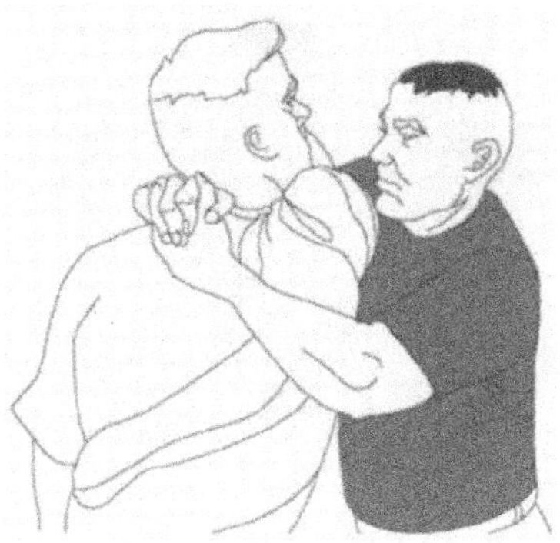

Pull the opponent toward the chest by pulling the clasped hands toward the chest.

Exert pressure on the side of the opponent's neck with the forearm.

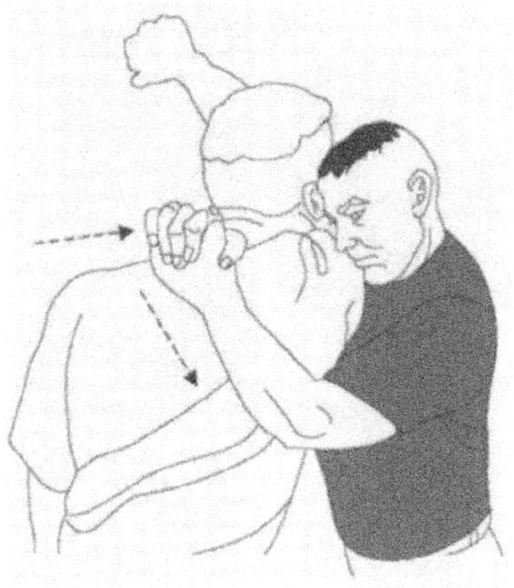

Rear Choke

Marines execute a rear choke when they are behind the opponent, the opponent is on the ground, or when they are taking the opponent to the ground. To execute the rear choke, Marines—

Reach, with the right arm, over the opponent's right shoulder and hook the bend of the arm around his neck.

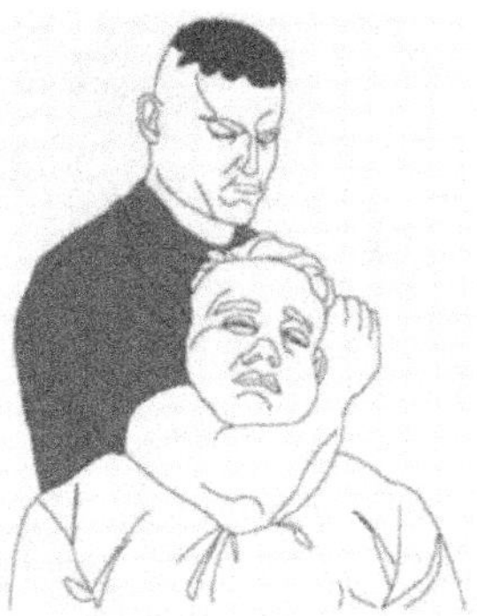

Clasp both hands together.

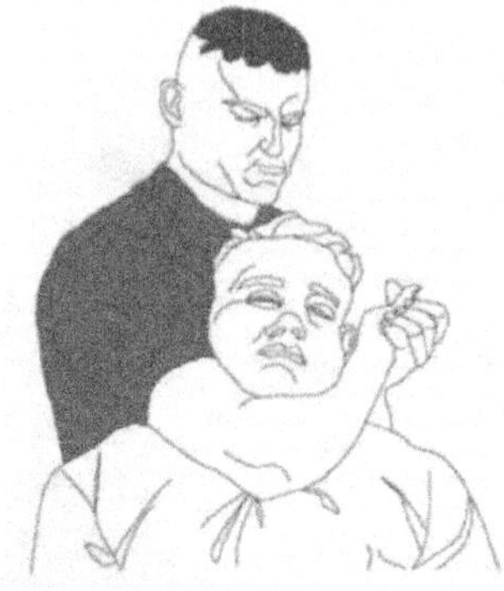

Exert pressure with the biceps and forearms on both sides of the opponent's neck on his carotid arteries.

Maintain pressure with the biceps and forearms on both sides of the neck and draw the opponent closer by drawing the right arm in.

Figure-Four Choke

A variation of the rear choke is the figure-four choke. The figure-four choke allows Marines to gain more leverage than the rear choke. If the rear choke cannot be secured, the figure-four variation is employed to increase the pressure of the choke on the opponent. To execute the figure-four variation of the rear choke, Marines—

Apply a rear choke. The Marine's body should be against the opponent's body.

Grasp the left biceps with the right hand and place the left hand against the back of the opponent's head.

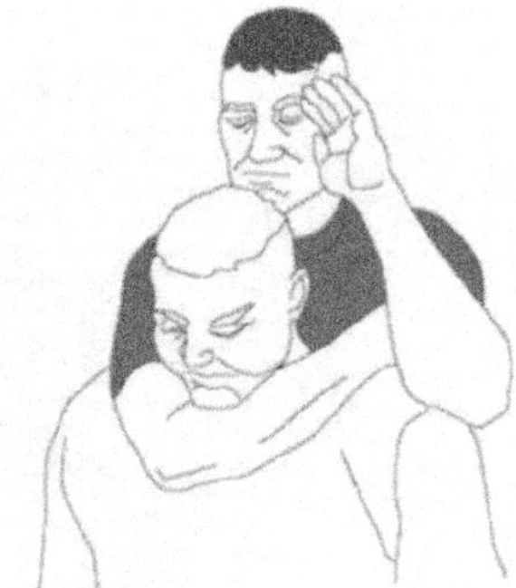

Push the opponent's head forward and down with the left hand.

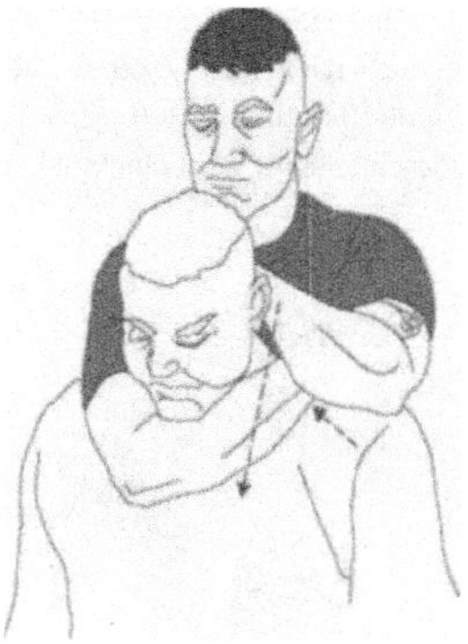

Draw the right arm in, maintaining pressure with the bicep and forearm on both sides of the opponent's neck.

3. Counters to Chokes and Holds

During a close combat situation, an opponent may apply a choke or hold on a Marine. If the opponent correctly applies a choke, a Marine quickly loses consciousness. If a choke is not executed properly, it often results in a hold, typically a bear hug or a headlock. A hold allows the opponent to control a Marine and removes the Marine's ability to attack. It is important for Marines to extract themselves from chokes and holds, regain the tactical advantage, and counter with strikes.

Although a choke causes unconsciousness in 8 to 13 seconds for a blood choke and 2 to 3 minutes for an air choke, the first movement in any counter to a choke is to clear the airway. Marines use softening techniques to loosen an opponent's grip and to clear their airway. Softening techniques are particularly effective if Marines lack the physical strength of their opponent. These techniques include groin strikes, eye gouges, foot stomps, etc. Softening techniques are not offensive; rather, they are used to loosen an opponent's hold.

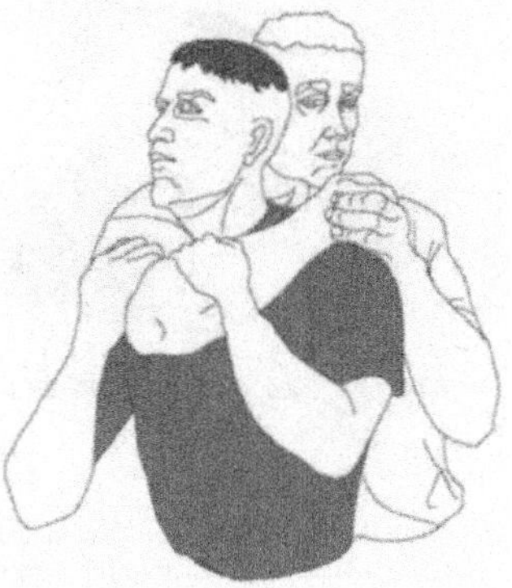

The second movement is to tuck the chin. Once the airway is clear, Marines tuck their chins to prevent the opponent from reapplying the choke.

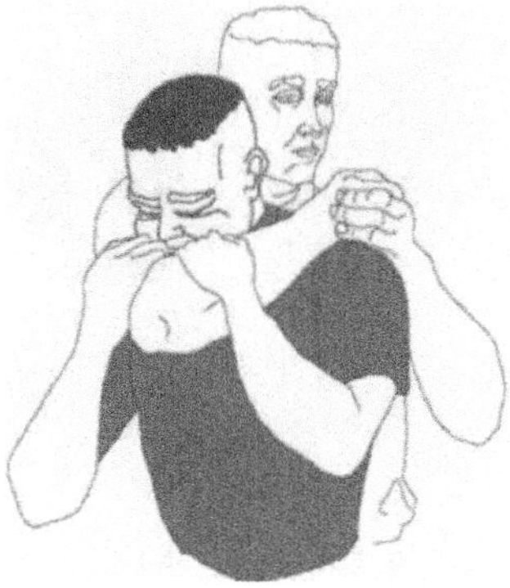

Counter to a Front Choke

Marines use a counter to a front choke when the opponent approaches from the front and uses both hands to choke a Marine around the throat. To execute the counter to the front choke, Marines—

Grasp the opponent's right forearm (where the elbow bends) with the left hand and apply downward pressure on the opponent's radial nerve with the fingers.

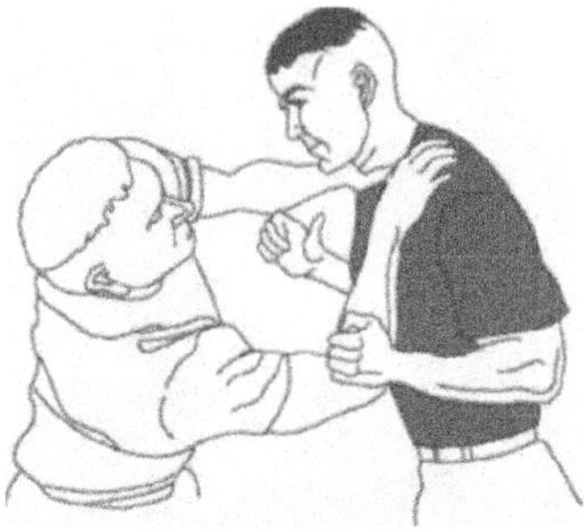

Execute a chin jab to the opponent's chin with the right hand. To generate power into the strike, bring the left foot to the outside of the opponent's right foot and rotate the hips into the strike.

Counter to a Rear Choke

Marines execute a counter to a rear choke when the opponent approaches from the rear and puts his right arm around a Marine's throat. To execute the counter to the rear choke, Marines-

Grasp the opponent's forearm (at the radial nerve) and bicep with both hands and pull down just enough to clear the airway. Once the airway is clear, tuck the chin to protect the airway and to prevent the opponent from reapplying the choke.

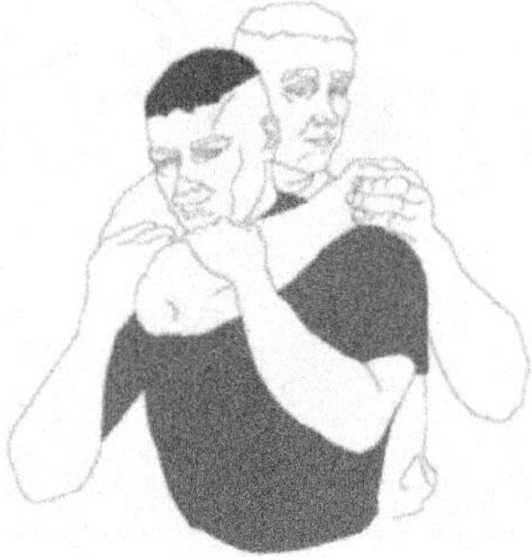

Step behind the opponent's right leg with the left foot, keeping both legs bent (almost in a squatting position).

Strike and drive the left elbow into the opponent's torso while rotating the hips and pivoting to the left, throwing the opponent back and to the ground.

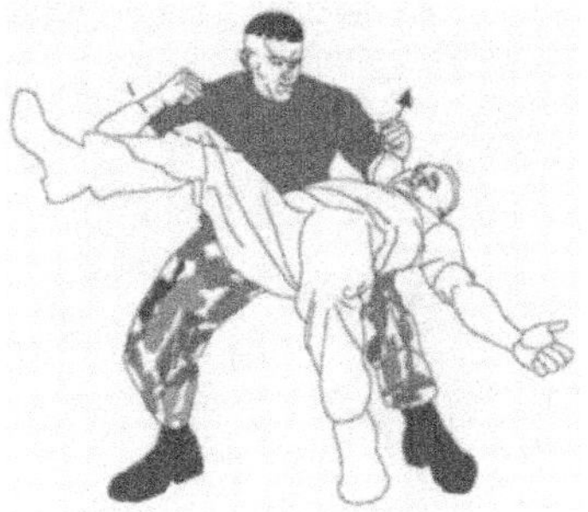

Counter to a Front Headlock

Marines use a counter to a front headlock when the opponent approaches from the front and puts his right arm around the Marine's neck, bends the Marine forward, and locks the Marine's head against his hip. To execute the counter to a front headlock, Marines—

Grasp the opponent's wrist and forearm with both hands and pull down to clear the airway. Maintain control of the opponent's wrist throughout the move. Once the airway is clear, tuck the chin to protect the airway and to prevent the opponent from reapplying the choke.

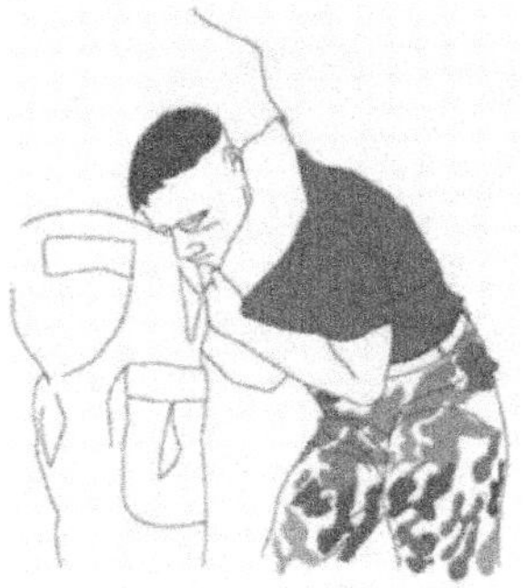

Move the right hand and arm across the opponent's torso.

Step forward and to the left with the left foot at a 45-degree angle.

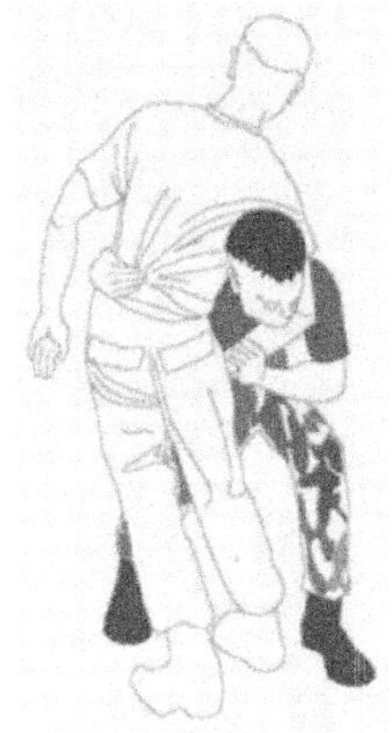

Execute a sweep with the right foot against the opponent's right leg. At the same time, push against the opponent's chest with the right arm and shoulder to generate power in the sweep.

Counter to a Rear Headlock

Marines use a counter to a rear headlock when the opponent approaches from the rear and puts his right arm around the Marine's neck, bends the Marine forward, and locks the Marine's head against his hip. To execute the counter to the rear headlock, Marines-

Grasp the opponent's wrist and forearm with the right hand and pull down to clear the airway. Once the airway is clear, tuck the chin to protect the airway and to prevent the opponent from re-applying the choke.

Reach over the opponent's right shoulder with the left arm.

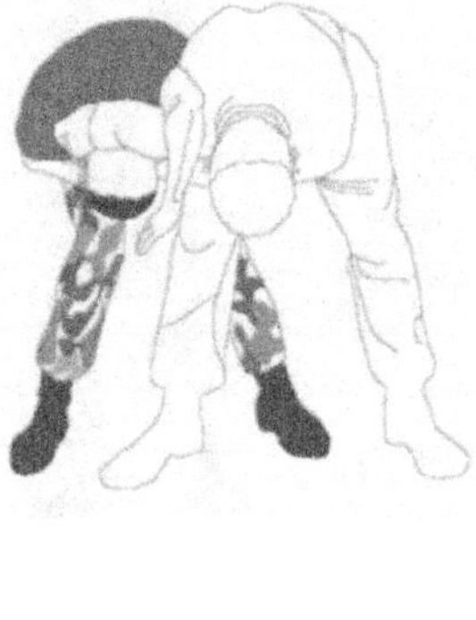

Grab any part of the opponent's face (chin, nose, eyes) and pull back while rising to a standing position.

Execute, with the right hand, a hammer fist strike to the opponent's exposed throat.

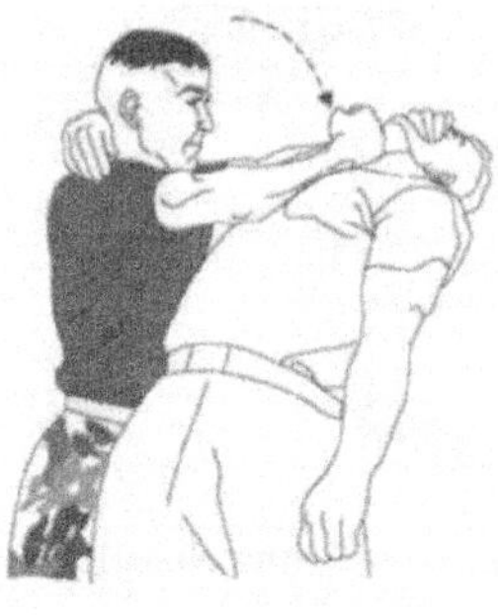

Counter to a Front Bear Hug

Marines execute a counter to a front bear hug when the opponent approaches from the front and puts both of his arms around the Marine's body, trapping the Marine's arms to the sides. To execute the counter to a front bear hug, Marines—

Step forward and to the left with the left foot at a 45-degree angle to the outside of the opponent's right leg, keeping the left leg bent.

Grasp the opponent's torso or arms to gain balance and to assist in throwing the opponent. It may be helpful to hook the opponent's right arm with the left arm.

Drive the right arm and shoulder forward and, at the same time, bring the right leg forward and sweep the opponent's right leg, bringing him to the ground.

Counter to a Rear Bear Hug

Marines use a counter to a rear bear hug when the opponent approaches from the rear and puts both of his arms around the Marine's body, trapping the Marine's arms to the sides. To execute the counter to a rear bear hug, Marines—

Step behind the opponent's right leg with the left foot, keeping both legs bent (almost in a squatting position). The left side of the body should be against the opponent's.

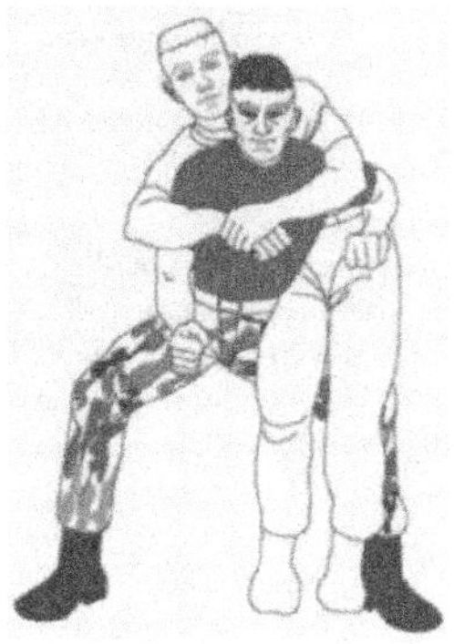

Pivot the hip, turning the body to the left and throwing the opponent back over the bent leg.

CHAPTER 7

GROUND FIGHTING

This chapter describes all techniques for a right-handed person. However, all techniques can be executed from either side.

In drawings, the Marine is depicted in woodland camouflage utilities; the opponent is depicted without camouflage. In photographs, the Marine is depicted in woodland camouflage utilities; the opponent is depicted in desert camouflage utilities.

Marines should avoid being on the ground during a close combat situation because the battlefield may be covered with debris and there is an increased risk of injury. However, many close combat situations involve fighting on the ground. The priority in a ground fight is for Marines to get back on their feet as quickly as possible. In any ground fighting scenario, Marines will usually end up in one of four positions with the opponent. The offensive positions, in which Marines have a tactical advantage, are the guard and mount. The defensive positions, which are used as counters when the opponent has the tactical advantage, are the counter to the guard and the counter to the mount. Marines can also employ chokes during ground fighting to quickly end a fight.

1. Offensive Ground Fighting

Guard Position

Marines execute a guard position when the opponent is on top and the legs are wrapped around the opponent's legs. If the opponent is on top, he may try to choke the Marine, but the Marine still has the tactical advantage because the Marine is in a position to control the situation. To execute the guard position technique, Marines—

Trap the opponent's hands on the chest by crossing the hands on the chest so the forearms are resting on the opponent's forearms. Apply pressure with the elbows or forearms.

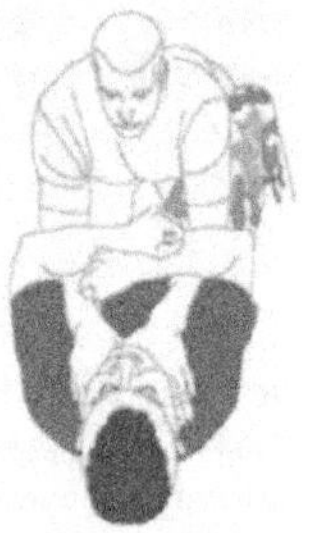

Strike the outside of the opponent's thigh with the cutting edge of the right heel. This causes the opponent to jerk to that side.

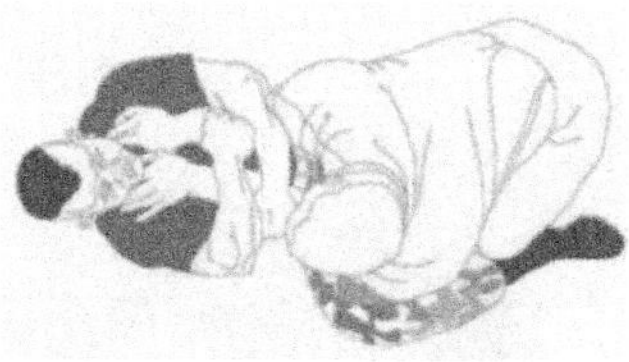

Move the head quickly to the left and swivel the hips to the right. At the same time, bring up both of the legs. Both legs are on the right side of the opponent's body.

Bring the right leg down, hooking the opponent's neck and head. Exert downward pressure to roll him over on his back. Grasp and maintain control of the opponent's left arm. Upon completion of the move, the Marine is sitting up with legs bent over the opponent while maintaining control of his arm.

Keep legs and knees bent. Maintain pressure against the opponent's neck with the back of the right foot and against his side with the left foot underneath his armpit. Squeeze the knees together, locking the opponent' s arm.

Pull the opponent's arm straight up and fall back sharply, pulling his arm to the side in the direction of his little finger.

Return to standing.

Mount Position

Marines execute the mount if the opponent is lying on his back on the ground and the Marine is on top with legs wrapped around the opponent's body. This position is an offensive position because Marines are in a better position to control the opponent and to execute ground fighting techniques. To execute the mount position technique, Marines—

Grab the opponent's wrists or forearms with the hands. Hold them tightly against the torso.

Maintain control of the opponent's arms with the left arm and apply pressure to the opponent's brachial plexus (tie in) with the right hand.

Bring the left foot close to the opponent's right armpit and swing the right leg across the opponent' s head.

Keep the legs and knees bent. Maintain pressure against the opponent's neck with the back of the right foot and against his side with the left foot which is underneath his armpit. Squeeze the knees together, locking the opponent's arm.

Pull the opponent's arm straight up and fall back sharply, pulling his arm to the side in the direction of his little finger.

Return to standing.

2. Defensive Ground Fighting

Counter to the Guard

Marines use this technique if the opponent is lying on his back on the ground and the Marine is kneeling on the ground between the opponent's legs. To execute the counter to the guard position, Marines—

Strike the opponent's femoral nerve, located on the inside of the thigh, with the elbows. This forces the opponent to separate his legs.

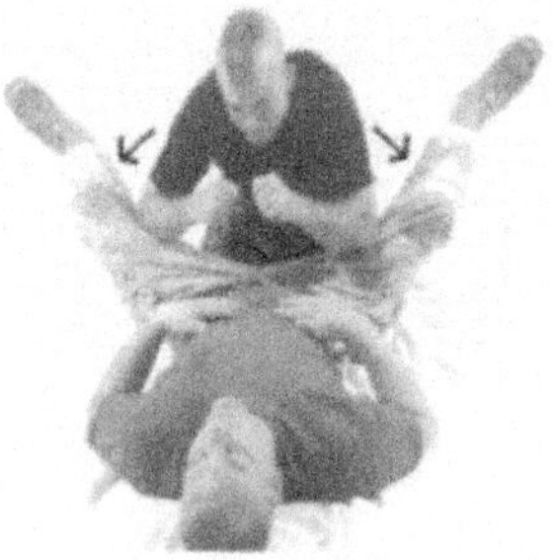

Strike the opponent's groin with the right fist.

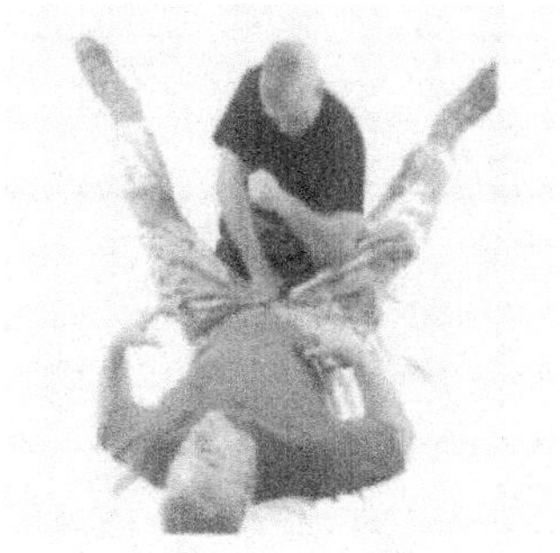

Hook the left arm underneath the opponent's right knee from the inside. Throw the opponent's leg over the head.

Duck the head quickly and move to the left.

Return to standing.

Counter to the Mount

Marines use this technique when they are lying on their back on the ground, the opponent is mounted on top of the Marine, and the opponent's legs are wrapped around the Marine. The opponent has the tactical advantage. To execute the counter to the mount position, Marines—

Grab the opponent's gear or clothing on his upper torso and pull him down close.

Use the right arm to hook the opponent's left arm, from the inside around the outside, above his elbow. With the right foot, hook the opponent's left leg or ankle.

Draw the elbow in to bend the opponent's elbow, bringing him down close. The arm must be hooked above the opponent's elbow in order to bend it.

Strike the opponent's side with the left hand.

Push the opponent over and roll him off to the right side.

Maintain control of the opponent's hooked arm and move to a kneeling position.

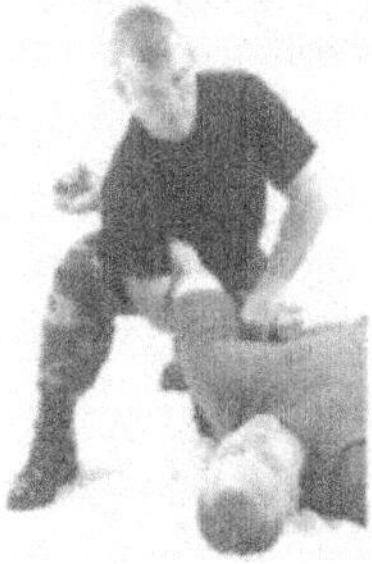

Move to a standing position while maintaining pressure on the opponent's arm and using the knee to apply pressure against the opponent's elbow.

3. Ground Fighting Chokes

WARNING

During training, never execute a choke at full force or full speed. Never hold a choke for more than 5 seconds.

The priority in ground fighting is for Marines to get back on their feet as quickly as possible. Sometimes, Marines can quickly end a ground fight by executing a choke on the opponent. When performed correctly, a choke can render an opponent unconscious in as little as 8 to 13 seconds. Chokes are easily performed regardless of size or gender. The chokes performed during ground fighting are the same as those performed while standing (see chap. 6).

Ground Fighting Front Choke

The ground fighting front choke is a blood choke performed most effectively from the mount position. The front choke employs the opponent's lapels or collar to execute the choke. To execute the ground fighting front choke, Marines—

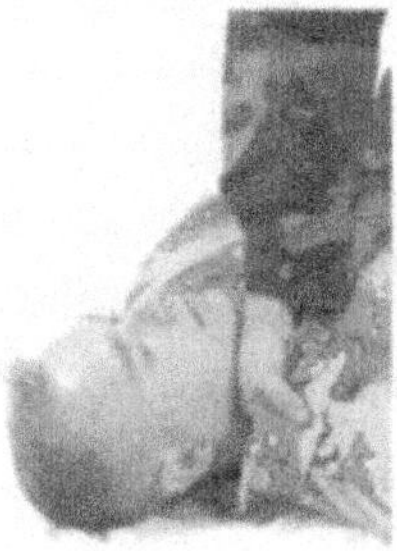

Grab the opponent's right lapel with the right hand, making certain that the knuckles or the back of the hand are against the carotid artery on the right side of the opponent's neck.

Keep the right hand pressed against the opponent's carotid artery, reach under the right arm with the left hand, and grab the opponent's left lapel, forming an X with the wrists.

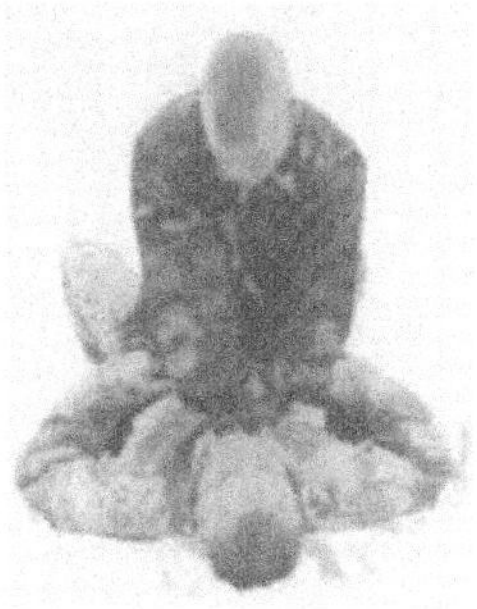

Keep the right hand pressed against the opponent's carotid artery and pull the opponent's left lapel to the left with the left hand.

Ground Fighting Side Choke

The ground fighting side choke is a blood choke performed from the mount position. The ground fighting side choke is particularly effective when the opponent raises his arms and places them on the Marine's chest or throat. To execute the ground fighting side choke, Marines-

Use the left hand to parry the opponent's right arm inboard (to the inside of the opponent's reach). Bring the right arm underneath the opponent's arm and up around the front of his neck.

Keep the fingers extended, place the back of the forearm against the opponent's neck just below his ear and press the carotid artery on his neck. Depending on the Marine's or the opponent's body configuration, the back of the thumb or wrist may press against the opponent's neck.

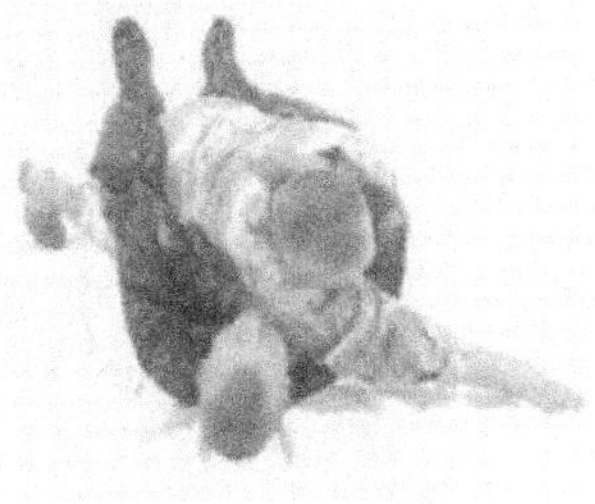

Use the left hand to reach around the back of the opponent's neck and clasp the hands together.

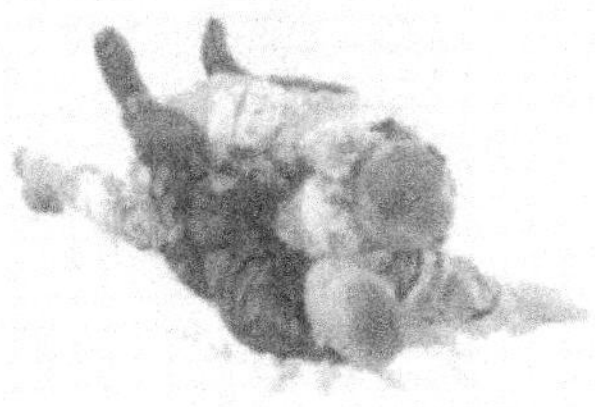

Pull the opponent toward the chest. Use the forearm to exert pressure on the side of his neck. This is done by pulling the clasped hands toward the chest.

Ground Fighting Rear Choke

The ground fighting rear choke is a blood choke performed when Marines are behind the opponent. To execute the ground fighting rear choke, Marines—

Wrap the legs around the opponent with the left foot against the inside of his left thigh and the right foot against the inside of his right thigh.

Use the right arm to reach over the opponent's right shoulder and hook the bend of the arm around his neck.

Use the left hand to clasp both hands together.

Exert pressure with the biceps and forearms on both sides of the opponent's neck. This applies pressure to the opponent's carotid arteries. While maintaining pressure with the biceps and forearms, draw the opponent closer by drawing the right arm in. To increase the effectiveness of the choke, lean backward by arching the back and pulling the opponent back. At the same time, push the feet against the opponent's thighs.

Figure-Four Choke

A variation of the ground fighting rear choke is the figure-four choke. The figure-four choke allows Marines to gain more leverage on the rear choke. If the rear choke cannot be secured, the figure-four variation may be applied to increase the pressure of the choke on the opponent. To execute the figure-four variation of the rear choke, Marines—

Apply the ground fighting rear choke. The chest should be against the opponent's back.

Grasp the left biceps with the right hand and place the left hand against the back of the opponent's head.

Use the left hand to push the opponent's head forward and down.

Draw the right arm in while maintaining pressure with the biceps and forearm on both sides of the opponent's neck.

Increase the effectiveness of the choke by arching backward and pushing the feet against the opponent's thighs. Continue to pull the opponent back with the right arm while exerting pressure forward with the left hand against the opponent's head.

CHAPTER 8

NONLETHAL TECHNIQUES

This chapter describes all techniques for a riglit-handed person. However, all techniques can be executed from either side.

In drawings, the Marine is depicted in woodland camouflage utilities; the opponent is depicted without camouflage. In photographs, the Marine is depicted in woodland camouflage utilities; the opponent is depicted in desert camouflage utilities.

The Marine Corps' involvement in military operations other than war—e.g., humanitarian, peacekeeping, or evacuation missions—has greatly increased. These missions require skills that span the spectrum of conflict and support operations within a continuum of force. But the Corps' day-to-day existence also demands a responsible use of force. Nonlethal techniques are among the skills Marines use to apply a responsible use of force.

1. Unarmed Restraints and Manipulation

Marines operate within a continuum of force, particularly in support of peacekeeping or humanitarian types of missions. In these situations, Marines must act responsibly to handle a situation without resorting to deadly force. Unarmed restraints and manipulation techniques including joint manipulation, come-alongs, and takedowns can be used to control a subject without resorting to deadly force. Marines must train to become proficient in nonlethal techniques and to respond in a responsible manner. These techniques are referred to as compliance techniques, and they are applied in the third level in the continuum of force.

WARNING

During training, never apply the techniques for unarmed restraints and manipulation at full force or full speed. Use a slow and steady pressure to avoid injury.

Compliance Techniques

Compliance techniques are unarmed restraint and manipulation techniques used to physically force a subject or opponent to comply. Compliance can be achieved through the close combat techniques of-

Pain compliance using joint manipulation and pressure points. (Pain compliance is the initiation of pain to get compliance on the part of the subject.)
Come-along holds.

Principles of Joint Manipulation

Joint manipulation is used to initiate pain compliance and gain control of a subject. It involves the application of pressure on thejoints (elbow, wrist, shoulder, knee, ankle, and fingers). Pressure is applied in two ways:

In the direction in which the joint will not bend. For example, joints such as the knees

and elbows only bend in one direction and when pressure is applied in the opposite direction, pain compliance can be achieved. Beyond the point where the joint stops naturally in its range of movement (i.e., it no longer bends).

Since each joint has a breaking point, Marines should apply slow steady pressure only until pain compliance is reached. Continued pressure will break the joint and may escalate the violence of the situation.

Joint manipulation also uses the principle of off-balancing. A subject can be better controlled when he is knocked off balance.

Wristlocks

A wristlock is a joint manipulation that can be applied in a number of ways to achieve pain compliance. The wrist rotates in a number of directions and will bend in a single direction until its movement stops naturally. In a wristlock, pressure is exerted beyond that point by bending or twisting the joint. A wristlock is executed when an opponent tries to grab Marines or is successful in grabbing Marines or their equipment. A wristlock can also be performed by Marines if they wish to initiate control of an opponent.

Basic Wristlock. A basic wristlock is executed when Marines grab the opponent's left hand with the right hand. To execute the basic wristlock, Marines—

Use the right hand to grab the opponent's left hand by placing the thumb on the back of the opponent's hand so that the Marine's knuckles are facing to the left.

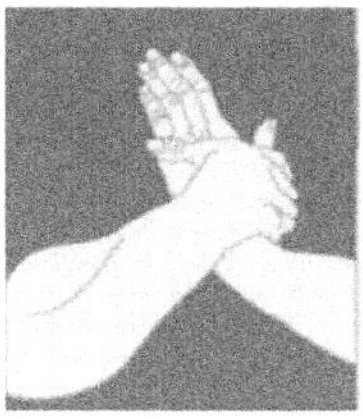

Hook the fingers across the fleshy part of the opponent's palm below the thumb. The fingers are used to anchor the hand so leverage can be applied to twist and bend the joint.

Exert downward pressure with the thumb to bend the opponcnt'sjoint. Rotate the opponent's hand to the right to twist the joint.

Step in to the opponent to keep the opponent's hand in close to the body to control him and provide more leverage on the wristlock.

Note: When executing the basic wristlock with the left hand, the Marine grabs the opponent's hand so that the Marine's knuckles are facing to the right, and then rotates and twists the opponent's hand to the left.

Use the left hand to further control the opponent.

Reverse Wristlock. A reverse wristlock is executed when Marines grab the opponent's right hand with the right hand. To execute the reverse wristlock, Marines—

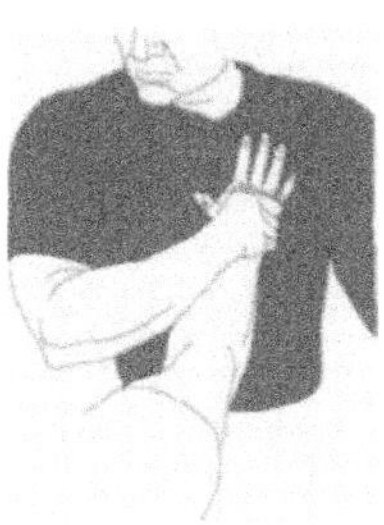

Place the right palm on the back of the opponent's right hand and wrap the fingers across the fleshy part of his palm below his little finger.

Twist the opponent's hand to the right while stepping in to place his hand against the chest. Apply downward pressure on the opponent's hand against the chest. Leave the opponent's hand on the chest to fully control the subject and to gain leverage.

Lean forward to use body weight to add additional pressure to the joint.

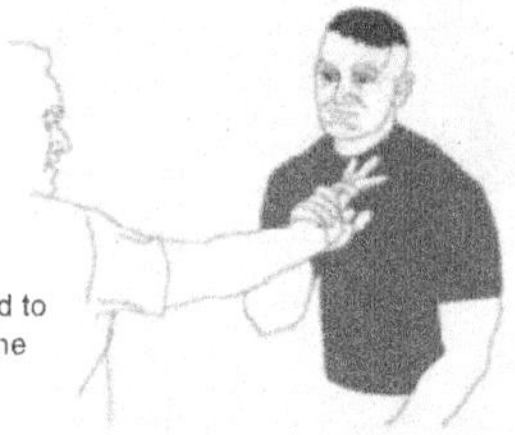

Use the left hand to further control the opponent.

Two-Handed Wristlock. Both hands can be used in the wristlock to maximize the leverage and pressure needed to bend and twist the joint. To execute the two-handed wristlock, Marines—

Place both thumbs on the back of the opponent's hand, thumbs crossed.

Hook the fingers of both hands around the fleshy part of the opponent's palm on both sides of his hand.

Step into the opponent and apply pressure downward on the back of his hand to bend the joint and rotate his wrist away from the body to twist the joint.

Enhanced Pain Compliance on Wristlock. Enhanced pain compliance techniques are applied in the third and fourth levels in the continuum of force. Additional pain be applied to a wristlock by-

Adding downward pressure to the elbow with the other hand or elbow by using the fingers to pull in on the opponent's radial nerve located on the inside of the forearm. When pressure is added to the opponent's radial nerve, his direction can be controlled.

Applying pressure against the opponent's finger joint to bend it in a direction it cannot bend (i.e., splitting the fingers).

Come-Along Holds

Marines use a come-along hold to control and move an opponent.

Escort Position. A common come-along hold is the escort position. To execute the escort position, Marines—

Face the opponent. Use the left foot to step forward at a 45-degree angle. Turn to face the right side of the opponent.

Use the right hand to firmly grasp the opponent's right wrist. With the left hand, firmly grasp the opponent's right triceps.

Position the opponent's controlled arm diagonally across the torso, keeping his wrist against the right hip. The Marine should be standing to the right of and behind the opponent.

Note: This technique works well when escorting an opponent on either the right or left side. Take caution when escorting an opponent by ensuring his controlled hand is not in a position to grab the holstered weapon. The preferred escort position is from the Marine's left side, so that the opponent is kept further away from the weapon.

Wristlock Come-Along. To execute the wristlock come-along, Marines-

Use the left hand to execute a basic wristlock. Incorporate the right hand in a two-handed wristlock for more control.

Maintain pressure on the opponent's wrist with the right hand, step forward, and pivot around to stand next to the opponent.

Release the left hand, quickly reach under the opponent's arm from behind, and grab his hand.

Use the left hand and apply downward pressure on the opponent's wrist.

Controlling Technique. The following controlling technique is used when an opponent grabs the Marine's wrist. To execute the technique, Marines—

Trap the opponent's hand with the palm of the other hand.

Rotate the opponent's trapped hand up and on his forearm while maintaining downward pressure on his trapped hand.

Apply downward pressure with both hands until the opponent is taken to the ground.

Armbars

An armbar is a joint manipulation in which pressure is applied on a locked elbow, just above the joint, in the direction the joint will not bend. An armbar has to be locked in quickly, but still requires a slow, steady pressure to gain compliance.

Basic Armbar. To execute a basic armbar, Marines-

Use the right hand to grab the opponent's right wrist.

Bring the left hand down on or above the opponent's elbow joint. To gain additional leverage, pivot to face the opponent.

Use the left hand to apply downward pressure on or above the opponent's elbow joint while pulling up on his wrist.

Armbar from a Wristlock. To execute an armbar from a wristlock, Marines-

Use the right hand to grab the opponent's right hand and execute a reverse wristlock.

Bring the left hand down on or above the opponent's elbow joint.

Use the left hand to apply downward pressure on or above the opponent's elbow joint while pulling up on his wrist.

Takedowns

A takedown is used to bring an opponent to the ground to further control him.

Takedown From a Wristlock Come-Along. To take the opponent to the ground from a wristlock come-along, Marines—

Use the right foot to push down on the opponent's calf or Achilles tendon.

Maintain control of the opponent's wrist and elbow and apply a slow, steady pressure to bring him to the ground.

Armbar to a Takedown. This technique is used to take a noncompliant opponent to the ground from an armbar. To execute the armbar takedown, Marines—

Use the right hand to execute a reverse wristlock.

Bring the left hand or forearm down on or above the opponent's elbow joint.

Pivot so the back is facing the opponent and, at the same time, lift the left elbow and slide the body so it is against the opponent, placing the armpit high above the opponent's elbow joint.

Lean back, placing the body weight on the opponent's arm until he complies or is taken to the ground.

Note: This technique may break the opponent's arm. Therefore, this technique should not be employed if the objective is a nonlethal takedown.

Wristlock Takedown. This technique is used to take a noncompliant opponent to the ground from a basic wristlock and to put him in a position where he can be handcuffed, if necessary. To execute the takedown, Marines-

Use the right hand to execute a basic wristlock. Incorporate the left hand in a two-handed wristlock.

Apply downward pressure on the wristlock, pivot on the ball of the right foot, and quickly turn to the right to take the opponent to the ground.

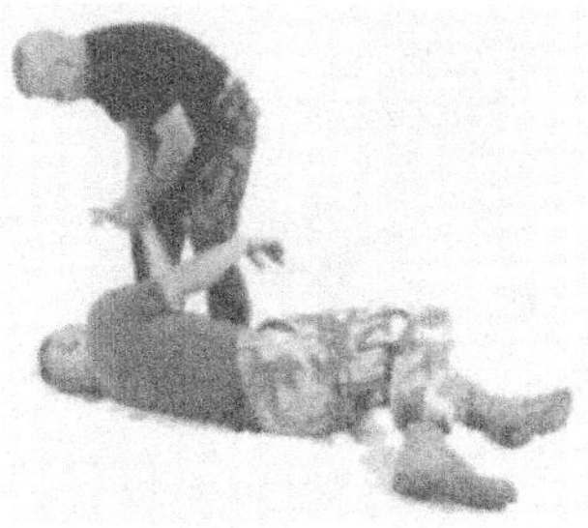

Continue to apply pressure on the wristjoint as the opponent lands on his back with his arm straight in the air.

Slide the left foot under the opponent's back, underneath his armpit.

Apply pressure with the knee against the opponent's triceps while pulling back on his arm.

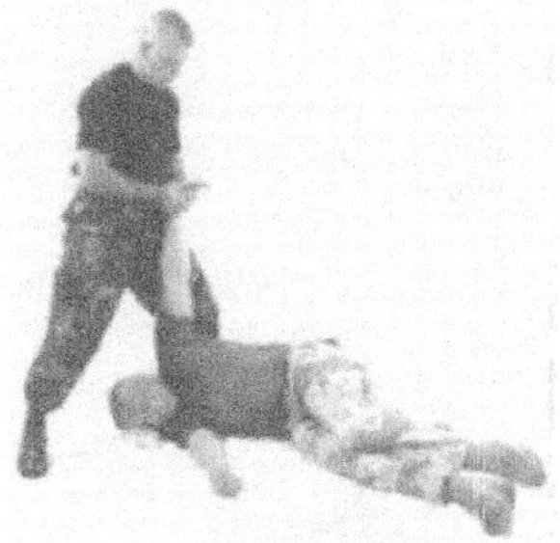

Continue applying pressure with the knee on the opponent's arm. Pivot and step around the opponent's arm to roll him on his stomach.

Kneel down with one knee on the opponent's back. The other knee is placed on the opponent's neck and shoulder on either side of his arm. Apply inward pressure with the knees to lock his arm in place.

Tell the opponent to put his other hand in the middle of his back. Bring the opponent's controlled hand to the center of his back.

Escort Position Takedown. This technique can be used to control a noncompliant opponent from the escort position. To execute the escort position takedown, Marines—

Lock the opponent's arm straight across the body while rotating his wrist away from the body.

Use the left hand or forearm and apply downward pressure above the opponent's elbow where the triceps meet.

Step back with the right foot and, keeping the opponent's hand controlled against the hip, pivot to the right while continuing to apply downward pressure on his arm to bring him to the ground.

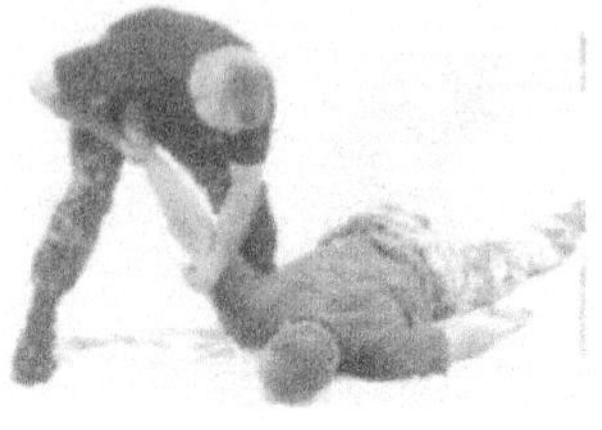

Note: This technique works well when escorting a subject on the Marine's right or left side. When taking down a subject from the right side, step back and pivot to the left.

2. Nonlethal Baton

A baton or nightstick can be an effective compliance tool when used correctly. Batons can be used defensively (blocking), offensively (striking), and as a restraining device when needed. In the fourth level of the continuum of force (assaultive [bodily harm]), defensive tactics include baton or nightstick blocks and blows. Blows to the head or other bony parts of the body are considered deadly force. When deadly force is not authorized, Marines must be able to employ blocks, strikes, and restraints with the baton with the minimum of force.

Grips

One-Handed Grip. To execute the one-handed grip, Marines—

Use the right hand to grasp the lower end of the baton, about 2 inches from the end.

Wrap the thumb and index finger around the baton so they are touching one another. The grip on the baton should be firm, but natural.

Two-Handed Grip. To execute the two-handed grip, Marines-

Use the right hand to grasp the lower end of the baton, about 2 inches from the end. Wrap the thumb and index finger around the baton so they touch one another.

Use the left hand to grasp the upper end of the baton, palm down, about 2 inches from the end. The hands should be approximately 10 to 12 inches apart.

Stance and Method of Carry

The basic warrior stance serves as the foundation for initiating nonlethal baton techniques. The method of cany provides effective defensive positions with a wide range of options to control a combative opponent.

One-Handed Carry. To execute the one-handed carry, Marines—

Grip the baton using the one-handed grip.

Elevate the baton, with the gripping hand at a level between the belt and shoulder.

Keep the left hand in the position of the basic warrior stance.

Two-Handed Carry. This carry is effective for blocks. To execute the two-handed carry, Marines-

Grip the baton using the two-handed grip.

Elevate the baton, with the left hand higher than the right hand.

Orient the weapon toward the subject.

Movement

In a nonlethal confrontation, movement may be made to create distance between Marines and the opponent or to close the gap to control him. When facing an opponent, Marines move in a 45-degree angle to either side of the opponent. Moving at a 45-degree angle is the best way to both avoid an opponent's strike and to put Marines in the best position to control the opponent.

Blocking Techniques

One-handed blocks are used when carrying the baton in the one-handed carry. The same one-handed blocks used in combative stick techniques (see page 3-6) apply in nonlethal baton engagements. The blocks include blocks for a vertical strike, a forward strike, and a reverse strike. Because the baton is often carried with two hands, there are also two-handed blocks that are effectively used from this carry. Two-handed blocks are discussed in the following subparagraphs.

High Block. Marines execute a high block to deter a downward vertical attack directed at the head and shoulders. To execute the high block, Marines—

Raise the baton up to a level even with or above the head. The baton should be in a horizontal position to block the blow. The fingers of the left hand should be open and behind the baton.

Place the baton perpendicular to the opponent's striking surface to absorb the impact of the blow.

Bend the elbows slightly to help absorb the impact of the blow. The arms should give with the strike of the blow.

Low Block. Marines execute a low block to deter an upward vertical attack directed at the abdomen, groin, or torso. The opponent's blow can be delivered by a foot, knee, or fist. To execute the low block, Marines—

Lower the baton to a level even with or below the groin. The baton should be in a horizontal position to block the blow. The fingers of the left hand should be open and behind the baton.

Place the baton perpendicular to the opponent's striking surface to absorb the impact of the blow.

Bend the elbows slightly to help absorb the impact of the blow. The arms should give with the strike of the blow.

Right Block. Marines execute a right block to deter an attack directed at the head, neck, flank, or hip. The opponent's blow can be delivered by a foot, knee, fist, or elbow. To execute the right block, Marines-

Thrust the baton in a vertical position to the right side.

Pivot to the right by stepping with the left foot and pivoting off the ball of the right foot. Rotate the hips and shoulders into the direction of the block. The fingers of the left hand should be open and behind the baton.

Place the baton perpendicular to the opponent's striking surface to absorb the impact of the blow.

Bend the elbows slightly to help absorb the impact of the blow. The arms should give with the strike of the blow.

Left Block. Marines execute a left block to deter an attack directed at the head, neck, flank, or hip. The opponent's blow can be delivered by a foot, knee, fist, or elbow. To execute the left block, Marines—

Thrust the baton in a vertical position to the left side.

Pivot to the left by stepping with the right foot and pivoting off the ball of the left foot. Rotate the hips and shoulders into the direction of the block. The fingers of the left hand should be open and behind the baton.

Place the baton perpendicular to the opponent's striking surface to absorb the impact of the blow.

Bend the elbows slightly to help absorb the impact of the blow. The arms should give with the strike of the blow.

Middle Block. Marines execute either a left or right block to deter an attack directed at the face, throat, chest, or abdomen. To execute the middle block, Marines—

Thrust the baton in a vertical position straight out in front of the body. The fingers of the left hand should be open and behind the baton.

Place the baton perpendicular to the opponent's striking surface to absorb the impact of the blow. The baton should be held with the left hand forward of the right.

Bend the elbows slightly to help absorb the impact of the blow. The arms should give with the strike of the blow.

Restraining Technique

The strong-side armlock is used to restrain an opponent who is not compliant. To execute the strong-side armlock, Marines—

Use the right hand to run the baton up under the opponent's left armpit, parallel to the ground.

Use the right foot to step forward at a 45-degree angle to the left side of the opponent. The baton is across his forearm.

Use the right hand to drive the baton forward and up so the action bends the opponent's arm behind his back. At the same time, continue moving around him to get behind him.

Place the baton on the opponent's forearm with the thumb and/or fingers and apply pressure to his forearm. At the same time, grasp the other end of the baton with the left hand.

Pull up on the low end of the baton with the right hand. At the same time, push down on the top end of the baton with the left forearm, reaching around with the left hand to grasp the opponent's biceps or shoulder.

Continue exerting downward pressure with the left forearm while pulling back on the opponent's biceps with the left hand. This places the opponent in a position where he is controlled and can be moved.

Apply pressure with the foot against the bend in the opponent's leg above his calf. This lowers the opponent to the ground rather than throwing him to the ground and risking severe injury.

Striking Target Areas

Marines must avoid striking an opponent in the head, neck, or other bony parts with the baton because this is considered deadly force and can lead to serious bodily injury or death. Instead, the legs, arms, and buttocks are target areas that are considered nonlethal.

Legs. Primary targets are the thighs and lower legs. Avoid striking the knee and ankle joints because this can cause permanent damage.

Arms. Primary targets are the upper arms. Avoid striking the shoulder and elbow and wrist joints because this can cause permanent damage.

Buttocks. Primary targets are the buttocks. Avoid striking any other part of the torso, including the chest, rib cage, spine, tail bone, and groin because strikes to these areas can cause permanent damage or death.

One-Handed Striking Techniques

One-Handed Forward Strike. A forward strike follows either a horizontal line or a downward diagonal line using a forehand stroke. To execute the one-handed forward strike, Marines-

Stand facing the opponent with the baton carried in a one-handed cany.

Place the right hand palm up, swing the baton from right to left, and make contact with the opponent.

One-Handed Reverse Strike. A reverse strike follows either a horizontal line or a downward diagonal line using a backhand stroke. To execute the one-handed reverse strike, Marines—

Stand facing the opponent with the baton carried in a one-handed carry.

Bend the right arm and cross the arm to the left side of the body. The baton should be close to or over the left shoulder.

Place the right hand palm down, swing the baton from left to right, and make contact with the opponent.

Two-Handed Striking Techniques

Two-Handed Forward Strike. This strike is an effective follow-up to a middle block or left block. To execute the two-handed forward strike, Marines—

Pull back with the left hand while driving the right hand forward toward the opponent. The baton should be horizontal to the ground. Power is generated by stepping forward with the right foot and rotating the right hip and shoulder into the strike.

Contact the opponent with the end of the baton.

Two-Handed Reverse Strike. This strike is an effective follow-up to a middle block or left block. To execute the two-handed reverse strike, Marines—

Pull back with the right hand while driving the left hand forward toward the opponent. The baton should be horizontal to the ground. Power is generated by stepping forward slightly with the left foot and rotating the left hip and shoulder into the strike.

Contact the subject with the end of the baton.

Front Jab. This strike is effective for countering a frontal attack. It can also be executed as a quick poke to keep a subject away. To execute the front jab, Marines thrust both hands forward in a quick jab. The baton is held either horizontal to the ground or at a slight downward angle.

Rear Jab. This strike is effective for countering a bear hug from the rear. To execute the rear jab, Marines thrust both hands rearward in a quick jab. The baton is held either horizontal to the ground or at a slight downward angle.

APPENDIX A

PUGIL STICK TRAINING

A pugil stick is a training device used to simulate a rifle bayonet so that effective, but safe, training can be conducted to build proficiency of rifle bayonet techniques. Pugil stick training builds on the techniques used to throw punches. Pugil stick training is the only "full contact" training provided to Marines in the Close Combat Program. Pugil stick training teaches Marines to function when faced with stress and violence, and it prepares them to deliver a blow and take a blow. It also provides them with the physical and mental skills vital to success on the battlefield.

1. Pugil Stick Training

Design

A pugil stick consists of a stick wrapped in padding at both ends that can be gripped like a rifle. The pugil stick is approximately the same weight and length of an unloaded rifle with a bayonet attached.

Holding the Pugil Stick

The pugil stick is held in the same manner as the sen ice rifle. All movements come from the basic warrior stance. To hold the pugil stick correctly, Marines—

Use the right hand to grasp the lower end of the pugil stick overhanded.
Use the left hand to grasp the upper end of the pugil stick underhanded.
Use the right forearm to lock the lower end of the pugil stick against the hip.
Orient the blade end of the pugil stick toward the opponent.

Safety Equipment

The following safety equipment must be worn during any pugil stick bout.

Groin Protection. Groin protection protects the groin from an accidental blow. It should be pulled high around the waist to protect the groin area, with the concave portion against the body.

Flak Jacket. The flak jacket provides protection to the body. It is worn completely fastened.

Neck Roll. The neck roll prevents whiplash if Marines receive a blow to the head. The neck roll further supports the head and protects the neck from blows. The neck roll is worn above the flak jacket and below the helmet. The tied end of the neck roll faces the front.

Helmet. A regulation football helmet protects the face and head. The helmet must fit snugly and the chin straps must be adjusted and snapped.

Mouthpiece. The mouthpiece is worn on the upper teeth to protect the teeth.

Gloves. Marines wear gloves to protect the hands if pugil sticks do not have gloves built-in.

Pugil Stick Screening

Prior to pugil stick training, instructors must ask participating Marines the following questions. If a Marine answers "yes" to any question, he must see the corpsman or a competent medical authority who evaluates him and determines whether or not the he can participate in the training.

- Are you on light duty?
- Are you restricted to running shoes by a corpsman or a doctor?
- Have you fought in a pugil stick bout within the last 7 days?
- Have you received a blow to the head within the last 7 days?
- Have you had a concussion within the last 6 months?
- Have you had dental surgery within the last 24 hours?
- Do you have stitches or staples on your body?
- Have you had a shoulder or head injury within the last 5 years?
- Are you taking a prescription drug?
- Do you have an ear infection or current sinus infection?
- Have you had a broken bone within the last 6 months?

Safety Personnel

The following safety personnel are required to conduct pugil stick training:

- One close combat instructor must officiate the bout.
- One close combat instructor trainer, commissioned officer, or staff noncommissioned officer must be in the training area to serve as range safety officer.
- One corpsman must be in the training area.

Training Area

To prevent injury, Marines train on areas with soft footing (i.e., sand or grass). Training mats are not recommended because feet can stick to the mats, prohibiting movement or causing joint injuries by twisting a knee. Bouts should not take place on a hard surface area; e.g., a flight deck or parking lot. A boxing ring may be used to conduct pugil stick bouts; ring dimensions can vary as long as there is ample room to execute the techniques in the training area.

Second Impact Syndrome

Second Impact Syndrome occurs when a second blow to the head produces a second concussion that occurs within 1 week following a previous concussion (before recovery from the first concussion). Second Impact Syndrome causes rapid brain swelling and can cause death. Therefore, there must be 7 days between pugil stick bouts to reduce the risk of severe injury resulting from Second Impact Syndrome. The 7 day separation between pugil stick bouts significantly reduces the possibility of injury, particularly in someone who may have suffered a brain injury or concussion but shows no symptoms.

Any Marine who experiences headaches or the following symptoms after training must be examined by appropriate medical personnel:

- Blurred vision.
- Ringing in the ears.
- Dilation of the pupils.
- Slurred speech.
- Bleeding from ears or mouth.
- Swelling in head or neck area.
- Any unnatural discoloration in head or neck area.

The Marine should not be allowed to participate in pugil stick training or any other activity where a heavy blow might be sustained for a minimum of 7 days after the headache or other symptoms have subsided.

Safety Measures

The following safety measures must be followed:

- Gear must be worn properly throughout training.
- Contact lenses or glasses will not be worn.
- False teeth will be removed from the mouth.
- Nothing will be worn around the neck except the neck roll.

Competition among groups of Marines is authorized as long as it does not overshadow training objectives or compromise safety procedures.

Safety and proper techniques are paramount.

Safety is more important than competition.

2. General Rules and Regulations Governing Pugil Stick Bouts

Instructors and Support Personnel Requirements

At a minimum, one close combat instructor observes a pugil stick bout. For safety purposes, it is better to have two instructors judging a bout because each instructor can fully observe each of the fighters. The best position for observation is to the right of a fighter. This allows the instructor to see the fighter's facial expression and body movement. The instructor's position must not interfere with the fight.

In addition, one commissioned or staff noncommissioned officer (to serve as range safety officer) and one corpsman will be in the training area.

Prior to the Bout

Prior to the bout, Marines—

Are paired according to height, weight, and gender.

Wear the proper safety equipment.

Once Marines have properly donned the safety equipment, they wait for instructions from the close combat instructor. On the command of the close combat instructor, a pair of Marines will enter the ring. Once in the ring, the close combat instructor inspects each Marine for the proper safety equipment.

Note: There should be enough gear so that when two Marines are training, two other Marines can be donning safety equipment for the next bout.

During the Bout

The bout begins when the close combat instructor blows the whistle. All strikes are directed above the waist. Upon hearing a whistle blast, all fighting immediately ceases.

Stopping a Bout

There are two reasons for stopping a bout: delivery of a scoring blow or an unsafe condition. The close combat instructor trainer, close combat instructor, or the range safety officer may stop the bout at any time an unsafe condition is observed.

Scoring Blow. A scoring blow is an offensive technique delivered to a vulnerable area of an opponent with sufficient force and precision to be considered as a disabling or killing blow. Scoring blows are not judged solely on the degree of force with which a blow is delivered, but on the accuracy and techniques employed. A scoring blow is defined as—

A straight thrust with the blade end of the weapon (red end of the pugil stick) to the opponent's face mask or throat.

A slash to the side of the opponent's helmet (below the ear) or neck with the red end of the pugil stick.

A heavy blow to the opponent's head with an authorized technique (i.e., buttstroke, smash) using the butt (black end) of the pugil stick.

When a scoring blow is delivered, the close combat instructor blows the whistle to stop the bout.

Unsafe Condition. The bout will be stopped as soon as an unsafe condition exists. An unsafe condition exists when a Marine is unable to defend himself, loses his balance and falls down on one or two knees or falls down completely, shows instability (e.g., buckling at the knees), loses muscular tension in his neck and his head snaps back or to one side, or appears disoriented. An unsafe condition also exists if a Marine lets go of one end of the pugil stick, the equipment (e.g., helmet, neck roll) falls off, or a Marine fails to use the proper techniques.

If any of these conditions occur—

The close combat instructor trainer, close combat instructor, range safety officer, or anyone supervising training stops the bout and separates the two Marines.

The corpsman evaluates the possibly injured Marine to determine if the fight can continue. The corpsman—

- Ensures that he is alert and responsive, both verbally and physically.
- Talks to him to see if he is coherent.
- Makes certain he comprehends and replies to verbal questions.
- Makes certain his speech is not slurred.
- Checks his physical signs.
- Ensures his eyes are focused and not dazed or glazed.
- Ensures his legs are not wobbly or shaky.

The pugil stick is not used as a baseball bat. The use of unauthorized techniques will result in expulsion from the bout. Only techniques taught in bayonet training are authorized. These techniques include-

Straight thrust.
Buttstroke (horizontal and vertical).
Smash.
Slash.
Parry.
Blocks (low, high, left, right).

After the Bout

A third whistle blast is used prior to identifying the Marine who delivered the killing blow.

A fourth whistle blast is used prior to demonstrating the killing technique used by the Marine who won the bout.

3. Directions for Making Pugil Sticks

Close combat instructor trainers are taught to make pugil sticks to standard specifications at school. It is recommended that all pugil sticks be made under the supervision of a close combat instructor trainer. Until the Marine Corps acquires a standardized pugil stick that units can order through the supply system, instructors use the following information and instructions to make pugil sticks locally.

Supplies

The following supplies are needed to make one pugil stick:

One 45" by 2" circular, oak, wood dowel.
Four 9" wide by 27" long foam pad pieces to make two inner handguards of double thickness.
Four 11" wide by 19" long foam pad pieces to make two outer handguards of double thickness.
Canvas material to cover inner and outer handguards.
One foam pad cut 1/4" thick by 7" wide by 13" long to make the centerguard.
One 8" wide by 14"long piece of canvas material to make a pugil stick sock.
Canvas material to make two pugil stick end caps. Ideally, two different color fabrics should be used to simulate the bayonet and buttstock ends of a rifle. Common colors for end caps are black and red. Each end cap is a different color to help the officiating certified close combat instructor determine the scoring blow during the pugil stick bout.
Two 1/4" thick by 11" wide by 54" long foam pad pieces to make two ends.
Two foam pads cut 1/4" thick by 3" wide by 14 1/2" long to make two end plug inserts.
One can of 3M Photo MountQ spray adhesive (NSN 8040-01-120-4009)to adhere foam padding to the stick. (Spray adhesive is available from self-service.)
One roll of duct tape or riggers tape (NSN 7510-00-266-5016) to protect foam pad and

canvas from wear and to reinforce gluing action of spray adhesive. (Duct tape or riggers tape is available from self-service.)

One roll para-cord or 5-50 cord to tie both the inner and outer handguards together.

One roll of thread strong enough to sew canvas material together.

Sixteen grommets, one for each of the four corners of the inner and outer handguards.

One foam sleeping mat (NSN 8465-01-109-3369). Use unserviceable foam sleeping mats obtained from Defense Reutilization and Marketing Office.

Tools

The following tools are needed to make a pugil stick:

Tape measure, ruler, or yardstick.
Scissors.
Grommet machine.
Knife.
Sewing machine.

Directions

To make a pugil stick, Marines perform the following steps.

Preparing Materials

Step 1. Cut foam sleeping mat into the dimensions listed above. Cut mats to double thickness.

Step 2. Use spray adhesive to adhere the two 9" by 27" pieces of foam together. Use the spray adhesive to adhere the two 11" by 19" pieces of foam together. Let adhesive dry before using the double-thick foam for handguards.

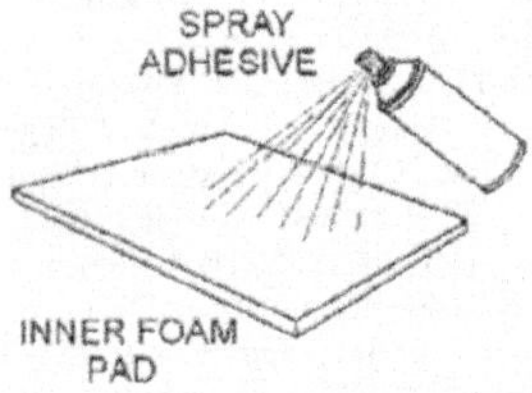

Step 3. Cut canvas material to make two coverings for the inner handguards. Cut material to make two coverings for the outer handguards. Two inner handguards and two outer handguards are required for each pugil stick.

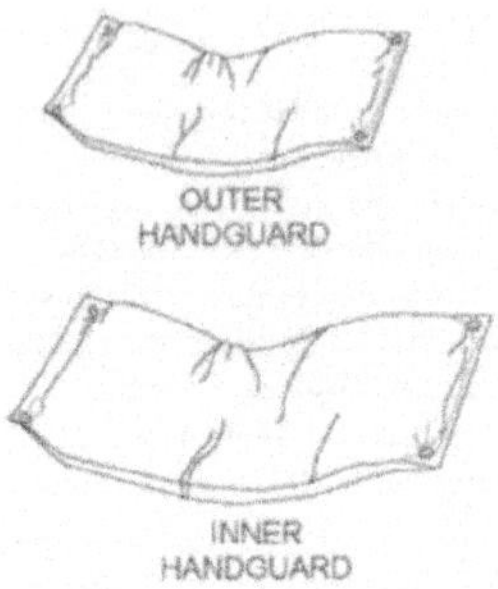

Step 4. Use a sewing machine to make a sleeve out of the 9" by 27" canvas. Sew down three sides. Insert foam pad inside the canvas sleeve and then sew down the fourth side. Use the grommet machine to create holes reinforced with metal grommets in each of the four corners. Once finished, the inner handguards are complete.

Note: It is recommended that the Fabric Repair Shop be used to sew canvas sleeves and apply grommets.

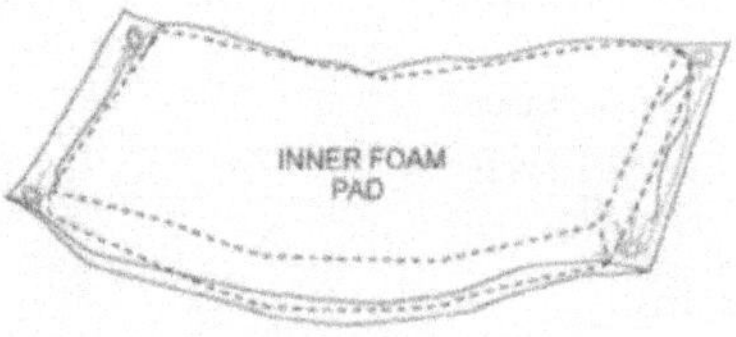

Step 5. Repeat step 4 to make the other inner handguard and both outer handguards.

Constructing Ends, End Plugs, and End Caps

Step 1. Apply spray adhesive on approximately 8 inches of one end of the dowel. This adhesive serves to glue the foam pad to the wooden dowel.

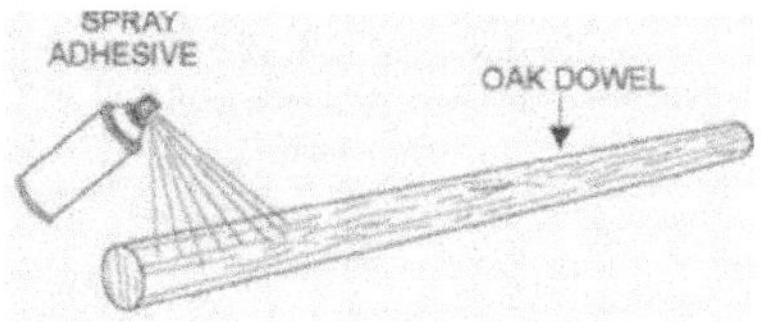

Step 2. Apply spray adhesive on one side of the 11" by 54" foam pad. This causes the foam pad to stick to itself when it is rolled around the end of the wooden dowel. Ensure 3 inches of the rolled foam pad extends past the wooden dowel for stability.

Step 3. Apply spray adhesive on one side of 3" by 14 1/2" foam pad. Roll pad into a tight cylinder.

Step 4. Apply spray adhesive on the exposed end of the dowel and fill the hole in the end with the end plug. The end plug prevents the wooden dowel from protruding beyond the edge of the rolled foam padding. This is an important safety precaution to prevent injury.

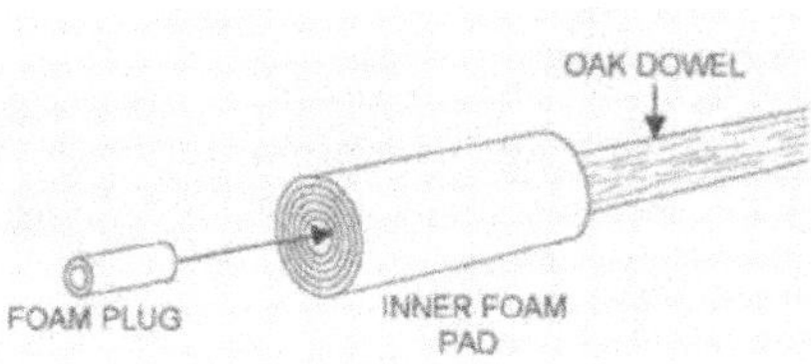

Step 5. Use duct tape or riggers tape to secure and reinforce the foam pad once the end plug is inserted in the end.

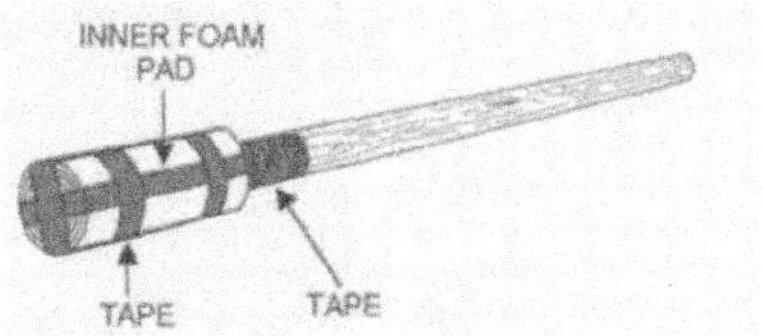

Step 6. Measure the dimensions of the ends (typically, the length of the cylinder should be 12 inches long to allow for taping). Cut and sew canvas material in the shape of a cylinder with one open end. Once complete, this becomes the end cap. The end cap is a colored sleeve that covers the rolled and taped foam pad at each end of the pugil stick.

Note: It is recommended that the Fabric Repair Shop be used to make the end cap.

Step 7. Slide one of the end caps over the rolled foam pad and tape it to the dowel.

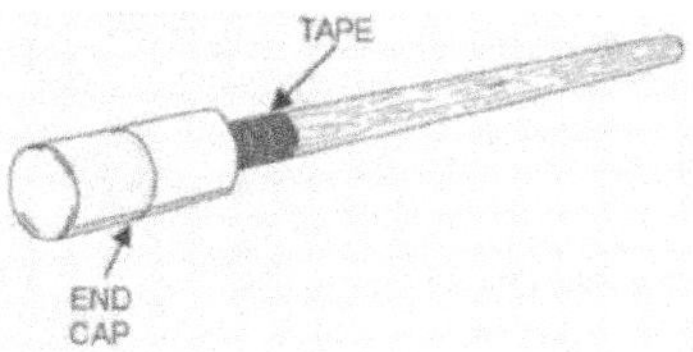

Building the Centerguard

Step 1. Spray one side of the 7" by 13" foam pad with spray adhesive and wrap it tightly around the center of the dowel.

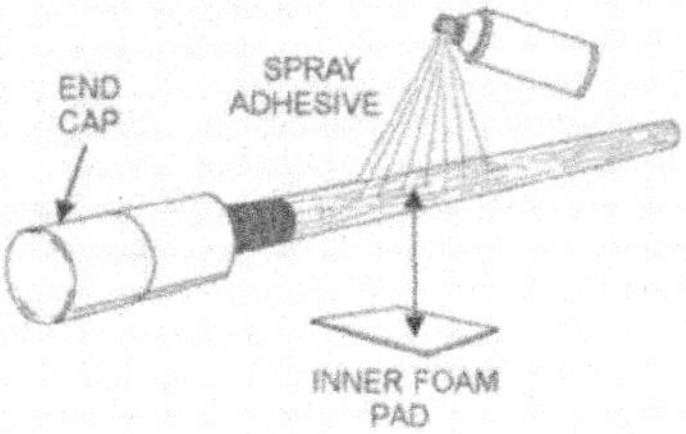

Step 2. Wrap duct tape or riggers tape around the foam pad to secure.

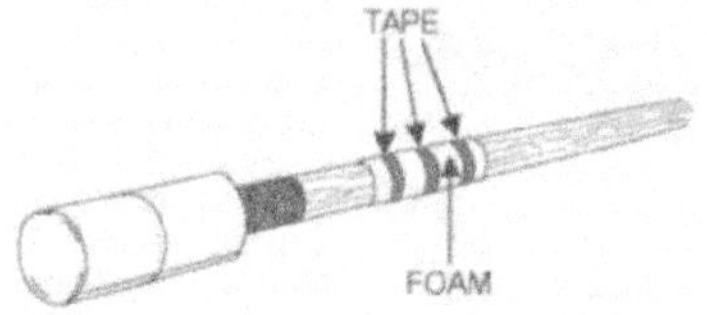

Step 3. Use a sewing machine to make a tubular pugil stick sock out of the 8" by 14" piece of canvas. Slide the pugil stick sock over the rolled foam pad and secure it with the duct tape or riggers tape. Duct tape or riggers tape protects the edges of the centerguard and reinforces the gluing effects of the spray adhesive.

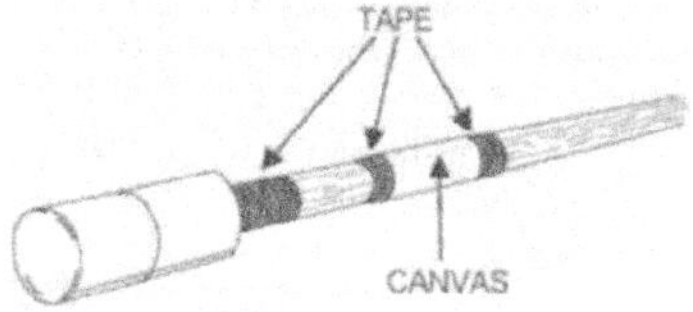

Note: It is recommended that the Fabric Repair Shop be used to make the pugil stick sock.

Finishing the Opposite End Cap

After one end cap is complete and the centerguard is finished, then repeat steps 1 through 7 in Constructing Ends, End Plugs, and End Caps for the other end of the pugil stick. Ensure that the opposite color end cap is used.

Fastening Handguards

Step 1. Place the inner handguards (longer piece) around the dowel. The inner handguard is placed on first because it protects the wrist. One inner handguard is needed for the left hand and one for the right hand.

Step 2. Fold the inner handguard around the pugil stick and run the para-cord or 5-50 cord through the metal grommets.

Step 3. Tighten down the corner ends of the inner handguards by pulling on the para-cord or 5-50 cord.

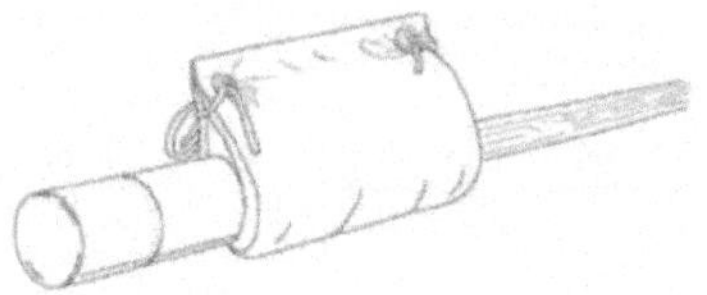

Step 4. Tie off the cord with a square knot above the metal grommets.

Step 5. Place the outer handguards (shorter piece) over the inner handguards (longer piece). Repeat steps 1-4 for the other handguard. The outer handguards provide additional protection to the knuckles and fingers and provides additional stability for the inner handguards.

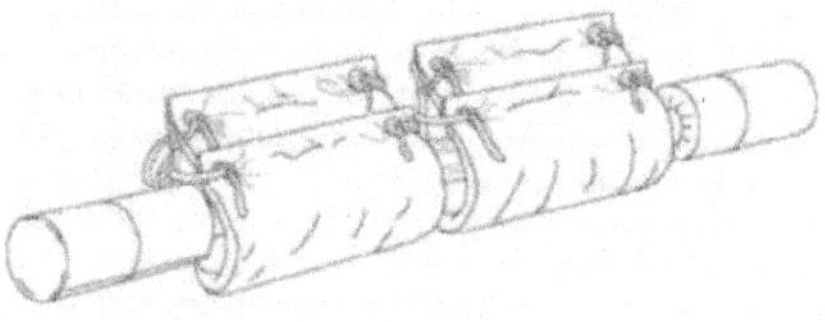

Serviceability Inspection

Prior to any training, all pugil sticks are inspected to ensure they are fit for use. If any of the following conditions exist, the pugil stick is unsafe and will not be used:

Stick protrudes through the end caps.
Frayed or worn edges exist on canvas surfaces.
The stick can be felt through the pads or the padding is too soft.
End caps or handguards are not securely attached.

APPENDIX B

SAFETY PRECAUTIONS DURING TRAINING

1. General Safety Precautions

When training close combat techniques, certain safety precautions must be adhered to to prevent injuries.

Most training should be conducted on a training area with soft footing such as a sandy or grassy area. If training mats are available, they should be used. A hard surface area is not appropriate for close combat training.

All techniques should be executed slowly at first. Marines can increase the speed of execution as they become more proficient. Marine-on-Marine training that requires contact (chokes, throws, ground fighting, and unarmed restraints and manipulation) should not be executed at full force or full speed.

If a technique is applied to the point that a Marine is uncomfortable, the Marine must "tap out." This indicates immediate release of the pressure being applied or to immediately stop the technique. The Marine "taps out" by firmly tapping his hand several times on any part of the opponent's body that will get his attention or by saying *stop*.

Second Impact Syndrome occurs when a second concussion develops within hours, days, or weeks following a prior concussion (and before recovery from the first concussion). Second Impact Syndrome causes rapid brain swelling and can cause death. Marines who experience headaches or other symptoms following training must be examined by appropriate medical personnel. These symptoms can include, but are not limited to, blurred vision, ringing in the ears, dialation of the pupils, bleeding from the ears or mouth, slurred speech, swelling in head or neck area, or any unnatural discoloration of head or neck. They should not be allowed to participate in pugil stick training or any other activity where a heavy blow might be sustained for a minimum of 7 days after the symptoms have subsided.

2. Safety Precautions for Individual Techniques

Falls

When training Marines to fall, they should progress from the ground, to a kneeling or squatting position, and then to a standing position. This ensures they are comfortable and understand the technique before progressing to executing falls from a higher profile. This instructional technique vastly reduces the risk of injury.

Strikes and Punches

When training Marines to strike and punch, they begin by executing the techniques "in the air." As they become more proficient, they execute strikes on equipment (when available) such as an air shield or a heavy bag. At no time should they be permitted to execute strikes on another student.

Chokes and Ground Fighting Chokes

When training Marines to execute chokes, they will not apply pressure to the opponent's throat during training because the trachea and windpipe can be crushed. During training, Marines should practice the proper procedures for blood chokes. They should never execute air chokes. At no time during training should a choke be applied at full force or full speed. No choke should be held for longer than 5 seconds.

Unarmed Restraints and Manipulation

When training Marines to execute unarmed restraints and manipulation techniques, they utilize slow and steady pressure. Never apply these techniques at full force or at full speed.

Pugil Stick Training

A blow to the head during training will bruise the brain. A second blow to the head can cause death. This is known as the Second Impact Syndrome. There must be 7 days between pugil stick training to prevent injury or death.

Introduction

Purpose

With United States involvement in Afghanistan, knowledge of the country is imperative. This "handbook" explains a number of basic issues that should be in the knowledge "toolbox" of a Marine working with Afghanis or deployed to the country.

Three Forces

Three main forces have shaped present day Afghanistan: Tribalism, Islam, and geography. They have permanent and far-reaching effects on Afghan society and on anyone involved with that society.

Geography

Even though it is a landlocked country, Afghanistan has been on the crossroads of history. The country and people have been at the mercy of numerous empires and conquering states. Afghanistan was a prize worth possessing for its river basins and fertile valley and for its location on the east-west trade routes, especially the Silk Route to China. This geographic factor has affected and changed the peoples of Afghanistan.

Ethnic Interaction

The interaction of the various ethnic groups that invaded the region contributed to the patchwork of people that make up present-day Afghanistan. Each of these external groups that invaded affected the inhabitants of what is now Afghanistan in unique ways.

- For example, Genghis Khan destroyed most of the towns in the thirteenth century, forcing much of population into nomadism.
- A later Turkic dynasty, The Timurids, reconstituted the empire, but was forced out by pressure from newer groups on the west, east, and north of the country.

This constant shift and pressure of various external and internal groups has molded the Afghan attitude toward outsiders and indigenous groups of the country.

Continued on next page

Tribalism

Tribalism is another key factor that has affected the social organization of Afghan society. It affects kinship, loyalties, honor, and even revenge.

[] The family as a unit can reach upwards of 50 people all united by the bond of a common "father."

[] This tribalism, with emphasis on the male or father's family, means that sons often bring their brides to live in their father's home.

Afghanistan remains a patrilineal society, in which membership in both family and lineage, as well as property, is inherited only by kin on the father's side of the family. Kin on the father's side of the family also form close alliances.

International "Tribalism"

This tribalism is not just a local issue. Tribalism is what holds much of the Osama bin Laden's Al Qaida (Al Qaeda) terrorist network together.

[] The bonding and the rules of social relations established in the village and nomadic setting play the same part in the international network.

[] This tribal set of values that applies in the farming and nomadic settlements of Afghanistan and Pakistan has the same validity in Bosnia, Macedonia, and the United States.

Islam

Islam is the last influence that affects Afghanistan in a permanent way. It is a basis for the daily living and for understanding how Afghanis approach interpersonal relationships. It is a guide to the moral and ethical values that Afghanis hold as well as their spiritual mindset.

Continued on next page

Table of Contents

The table below lists the main topics and their page numbers.

Quick Facts on Afghanistan

Introduction

This brief resource guide covers a number of important subjects that give you a clearer understanding of the country. This page of quick facts is like a job aid. It highlights critical issues.

Land of Contrasts

Afghanistan consists of a major mountain range, the Hindu Kush, with its valleys, deserts and river system, http://fermi.jhuapl.edu/maps/afghanistan/ This territory includes many varied geographical and ecological zones. The different peoples of Afghanistan are as distinct and varied as the geography. Bringing them under one flag has made them into a "single" society.

Name

Transitional Islamic State of Afghanistan (TISA) as of December 2002 (formerly called the Islamic Emirate of Afghanistan by the Taliban in 1997) is a land-locked country in southwest Asia.

Size and Population

Area: 652,225 square kilometers or 251, 772 square miles. Estimated population: @ 27,000,000.

Population; Life Expectancy

0-14 years: 42.2%; 15-64 years: 55.01%; 65 years and over: 2.79%.
Life expectancy about 46 years.

Languages and Ethnicity

Pushtu 35%, Afghan Persian (Dari) 50%, Turkic languages (primarily Uzbek and Turkmen) 11 %, 30 minor languages 4%.

Religion

Sunni Muslim 84%, Shi'ite (Shi'a) Muslim 15%, other 1%:.

Administrative Sectors

Country is divided into 30 provinces or velayets. The largest in area are Herat, Helmand, Qandahar (Kandahar), and Farah. The largest in terms of population are Kabul, Herat, Balkh, and Ghazni.

Population Centers

Principal towns include Kabul, Kandahar (Qandahar), Herat, Mazar-i-Sharif, Jalalabad, and Ghazni.
http://www.afghana.com/GetLocal/Afghanistan/Provinces.htm

Continued on next page

Quick Facts on Afghanistan, Continued

Government/ Legal System	A transitional government, consisting of a grand assembly, president, and judiciary; no longer based on the Qur'an (Koran), the holy book of Islam.
Politics	Taliban (religious students) headed by Mullah Mohammad Omar was deposed in late 2001; a transitional government based on United National Islamic Front for the Salvation of Afghanistan or UNIFSA, often referred to as the Northern Alliance, and exiled leaders, forming a legislative assembly, a presidency, and a judiciary.
Economic Sectors	Agriculture: 53% of gross national product (GNP); industry: 28.5% of GNP; services: 18.5% of GNP (1990).
Additional Sites	http://www.odci.gov/cia/publications/factbook/gcos/af.html — CIA site that has handbooks of countries of the world. http://ebooks.whsmithonline.co.uk/htmldata/encv.asp?mainpage=HTTP://EBOOKS.WHSMITHONLINE.CO.UK/ENCY CLOPEDIA/01/C0000001. HTM — short summary about Afghanistan from *The Hutchinson Family Encyclopedia.* http://www.nutshellnotes.com/afghanistan text.htm - short dictionary-like definition information about the country.

Geography and Climate

Land-locked

Afghanistan is a landlocked country in southwest Asia. To the north, its neighbors are Turkmenistan, Uzbekistan, and Tajikistan. Iran is to the west; the People's Republic of China to the northeast, and Pakistan to the east and south, http://www.lib.utexas.edu/maps/cia01/afghanistan_sm01.jpg

Chief Topographic Feature

The Hindu Kush mountains are the main topographic feature that marks Afghanistan. These mountains are a barrier between the northern provinces and the rest of Afghanistan. They divide the country into three distinct geographic areas: the central highlands, the northern plains, and the southwestern plateau. You can find good maps of Afghanistan at http://www.lib.utexas.edu/maps/middle_east_and_asia/afghanistan_topo86.jpg

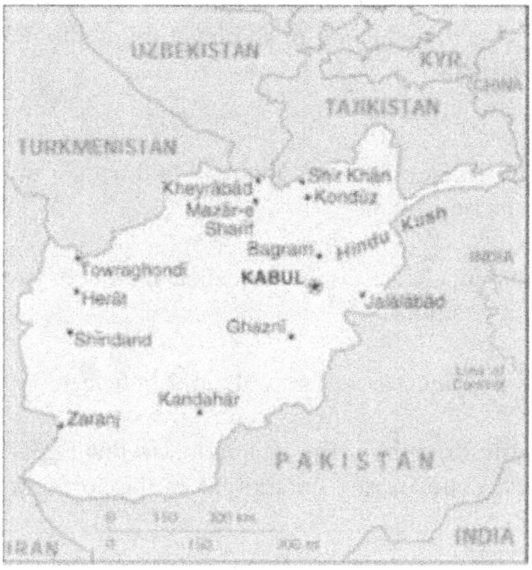

Source: http://www.globalspecops.com/oef.html

Continued on next page

Geography and Climate, Continued

Central Highlands

The central highlands are an area of about 160,000 square miles. This part of Afghanistan has deep, narrow valleys, as well as high mountains, which have proven to be historically important to the defense of the country. The Khyber Pass, one of the most famous routes to the Indian subcontinent, is located in the mountain ranges of the central highlands, http://www.maps.com/cgi-bin/search/hyperseek.cgi?search=CAT&Calegory=Asia%3AAfghanislanP&0ualifier=

The climate of the central highland is usually dry, with temperatures in the summer averaging around 80 degrees Fahrenheit; the winters are very cold.

Southern Plateau

This region of Afghanistan consists of high plateaus and sandy deserts. This desolate region covers about 50,000 square miles and is crossed by several large rivers. The average altitude of the southern plateau is about 3,000 feet. Kandahar (Qandahar), at an elevation of about 3,500 feet, enjoys a dry, yet mild climate.

Northern Plains

The northern plains are about 40,000 square miles of extremely fertile foothills and plains.

fj The Amu River (formerly called the Oxus River) runs thought the foothills of this region.

• The average elevation is about 2,000 feet.

Agriculture is the main occupation in this region. There are large amounts of mineral deposits and natural gas in the northern plains.

Climate

The climate of Afghanistan varies dramatically between the mountain country and the lowlands. In the southwest during summer the temperature can reach 120 degrees Fahrenheit. In the Hindu Kush mountains during winter it can fall as low as -15 degrees Fahrenheit.

Map Site

http://www.lib.utexas.edu/maps/afghanistan.html — This site has many different kinds of maps of Afghanistan.

Economy and Livelihood

Economic Areas

[] Most people in Afghanistan are farmers or peasants who cultivate crops for personal use or for marketing and selling as produce. Herding is the next favored occupation by population size, followed by forestry and fishing.

[] Manufacturing is the second largest area, followed by wholesale and retail commerce, then by transportation and mining. (With the current chaotic situation in the country, much of the economy has been disrupted.)

Agricultural Economy

Agriculture requires five major means of production: land, water, seed, labor, and animals (or tractors). Water is the most important resource in this respect. Each village has an official chosen by the elders or the village council. He regulates the division of water among landowners.

Family Farming

The family farm is the most common form of ownership and land cultivation. The family usually provides the labor and animals to cultivate its own land. Other agricultural labor arrangements include:

[] Sharecropping, where the tenant receives a part of the produce.
[] Commercial fanning, where the owner pays cash to the farmers.

Because of the current chaotic conditions in Afghanistan, these agricultural systems are in decline.

Herding

Pastoral nomadism has been the traditional way that Afghanis have raised and herded their cattle, sheep, and goats. These nomads usually maintain their herds in groups, traditionally as clans or tribes. They might rely on the peasants or farmers for staples such as grain and implements they need for running a household.

Continued on next page

Mobility: An Advantage

Because of their mobility nomads were:

[] The bearers of information and news.

[] A traditional source of credit because their herds were a "countable" tangible source of cash. In the towns and villages where they took their stock to sell, nomads could buy goods on credit to sell to the isolated farmers or peasants living near their summer pastures.

[] Often smugglers, both inside the country and across the borders.

Again, based on the chaotic war conditions of the country the usual nomadic herding patterns may have diminished or disappeared.

Manufacturing and Commerce

Besides food staples, such as margarine and wheat flour, woven fabrics, clothes and shoes are produced.

[] Fertilizer, cement, and electric energy are the three main industrial products.

[] Mining includes hard coal, gypsum, and natural gas.

The war in Afghanistan has caused industrial and commercial output to decline.

Languages and Ethnicity: Diversity

Introduction

As a landlocked nation, Afghanistan is at the crossroads for much migration and many settlements. The result of this exposure is a country with diverse ethnic and linguistics (language) groups.

Two Perspectives

[] The next two sections give you an idea of the kinds of people living in Afghanistan, including the ethnic groups of the country. An ethnic group consists of people who have many cultural traits in common. These may include language, religion, morals, customs and ceremonies, economics and livelihood.

[] Another way to look at diversity in a country is according to the languages the peoples of the country speak. Languages may cross ethnic boundaries, for example in Afghanistan, Dari is a common language to many of the different ethnic groups in the country. It crosses the ethnic/cultural boundaries of these different groups.

Languages of Afghanistan

Pushtu (Pashtu)	Pushtu is the official language of Afghanistan as, for example, Turkish is the official language of Turkey. Pushtu is also spoken in northern Pakistan. The language is written in a modified Arabic script.
Dari	Dari or Afghan Persian or Eastern Farsi is spoken by about 43% of the population. It is often referred to as the language of business and commerce.
Turkic Languages	About 11% of the population speaks a Turkic language, mainly Uzbek or Turkmen. Afghanis living close to the border of Turkmenistan and Uzbekistan probably speak Turkic languages.

Ethnic Groups: Pashtun

Introduction

Like the languages of the country, there are various ethnic groups in Afghanistan. The main groups are the Pashtuns (pronounced PAH-shtoons) (Pathans), Uzbeks, Hazaras (Hezoras), Tajiks (Tadjiks), and other minor groups.
http://www.lib.utexas.edu maps/middle east and asia/afghanistan ethnoling 97.jpg

Definition

The Pashtuns live in southern and eastern Afghanistan in a population belt stretching from Kabul in the east to Kandahar (Qandahar) to Zaranj, which is on the border with Iran. They also live in western Pakistan.

[] Pashtuns speak Pushtu (Pashtu) and are Sunni Muslims (see the section on **Islam** for an explanation of Sunnis). Many Pashtuns also speak Dari, the second language of Afghanistan.

[] Pashtuns make up about 38% of the population of Afghanistan. Until the recent Taliban take-over, the Pashtuns controlled the ruling dynasty of the country for about 200 years.

Settlements

Pashtuns generally live in tribal villages. A number of villages may be near a town. A village is built primarily with regard to the location of water and to self-defense. The dominant lineages or extended families usually have the best and most strategic lands. Familial ties determine a group's location relative to the dominant tribe.

Nomads

Some Pashtuns are herders. They move from site to site following the pastures as the seasons change. Like the permanent villages mentioned above, the campsites of these nomadic Pashtuns are based on tribal lineages. Their tents are made from goat hair, supported by posts or arched poles.

Economy and Livelihood

Pashtuns work at grain farming and animal husbandry (see **Nomads** above). Wheat is the most important crop. Other important products are fruits, nuts, vegetables, and opium.

Industry usually includes trades such as shoemaking and carpentry. Some Pashtuns are small businessmen and traders.

Continued on next page

Kinship

The tribe is the unit of kinship. Descent is claimed from the father's side of the family.

[] The clan is the principal division of a larger tribe. A clan is usually the offspring of one man, often including four or five generations, what is sometimes referred to as an extended family in Western terms. In this clan, a person marries and forms bonds.

[] The smallest kinship unit is a household. The household or *kor* is the main economic and social unit; its members may form a village, single housing compound, or a nomadic group.

Local Authority

Pashtuns have a concept and practice of local authority based on tribalism.

[] Honor *(nang)* is a tradition that pervades Pashtun society. It means there is little social class and no central authority.

[] Taxes or rent *(qualang)* gives status and social class to the local people, many of whom are landowners. Qualang also means that there may be a "centralized" authority at the local level.

National Government

Generally, villages consider that the national government has responsibility for conscripting young men into the military and for taxation, both of which take resources away from the clan/tribe. Thus, villagers may have an attitude of neutrality toward, non-cooperation with, or defensiveness against the central government.

Important Customs

Pashtuns generally believe in and follow these customs:

[] Purdah: Strict separation of the sexes in public
[] Chadori: Women wearing the veil.
[] Pakhtunwal: A mixture of blood-line revenge, hospitality to guests, defense of those who are in one's care, chastity of married women and caring for the weak

Ethnic Groups: Tajik (Tadjik)

Definition

Tajiks (pronounced TAH-jeeks) make up about 25% of the population of Afghanistan. They live

[] In a large area north and west of Kabul (this area borders on the Republic of Tajikistan)
[] Clustered around the city of Herat in western Afghanistan

Tajiks are also the principal inhabitants of the Republic of Tajikistan, a former region of the Soviet Union.
hltp://news.bbc.co.uk hi english static in depth world 2001 war on terror key maps ethnic groups.stm

Language and Culture

Tajiks are primarily Sunni Muslims. They speak the Tajik and Dan languages. Tajiks are believed to be descendants of the ancient Persians.

Economy and Livelihood

Since Tajiks in urban areas are well educated and wealthy, they make up the elite of Afghanistan. Their influence is in commerce.

[] The plains-dwelling Tajiks live mainly in Herat, Parwan, and also around Kabul. They are town dwelling and are traders, skilled artisans, and farmers, all middle-class. Because they have settled in towns, they have replaced their tribalism with an urban orientation and a strong sense of community loyalty.

[] Tajiks in rural areas make their living by fanning and herding. They are a sedentary people. Landowners are usually village leaders.

Politics: Importance in Northern Alliance

Tajiks are the second largest group after the Pashtuns. They are also the Pashtuns' closest rivals for power and prestige, but with two exceptions, one in the 14th century and one for 9 months in 1929; the Tajiks have never ruled their region.

Tajiks fonned an important part of the Northern Alliance, which opposed the Taliban. From a group of Tajik farmers, the chief leader of the Northern Alliance, Ahmad Shah Masoud, emerged. First he fought a guerilla war against the Soviet Communists and later against the Taliban. Taliban supporters assassinated him in late summer 2001.

Ethnic Groups: Hazara (Hezora)

Definition

Hazaras (pronounced HAH-zahrahs) make up about 19% of the Afghani population. They are Shi'ite Muslims, not Sunnis (for more details on Shi'ites, see the section entitled **Islam).**
http:;/news.bbc.co.uk hi english static in depth world 2001/war on terror'key maps /ethnic groups.stm

Groups/Settlements

The Hazaras fall into three groups: Settlements are located in an areas bounded

[] On the north by the town of Bamian (Bamyan)
[] On the west by Chaghcharan
[] In the east by Kabul
[] On the southeast by the town of Ghazni

The Hazaras themselves call this area Hazaristan, country of the Hazaras. Many Hazaras were forced off their land by the expansion of the majority Pashtuns and now live in Kabul and other towns.

Language

The Hazaras speak a language called Hazaragi. It is a mixture of Farsi (the language of Iran), Mongol, and Turkish.

Religion

[] Because they are Shi'ites, the Hazaras have been subjected to oppression over the years. Most recently, the Taliban massacred about 300 Hazara men in January 2001. The reason was the Hazaras were suspected of collaborating with the United Front (Northern Alliance).

[] Since they are Shi'ites, the Hazaras get support from Iran, which is predominantly a Shi'ite Muslim country.

http://news.bbc.co.uk/'hi/english uorld south asia newsid 1500000 1500682.stm — discusses Taliban persecution of Hazaras.

Continued on next page

Ethnic Groups: Hazara (Hezora), Continued

Origins

The origin of the Hazaras is interesting. Some researchers believe that Haz.aras are the descendents of the 12th century Mongol warrior, Genghis Khan and his followers; others trace the Hazaras to the Buddhist civilization of 2000 years ago, the Bamiyan, whose center was in Afghanistan.

http: // www .hazara. net/hazara/geography/B uddha/buddha.html — an article about relationship of Hazaras to the Buddhist statues.

Economy and Livelihood

The primary occupations of the Hazaras are fanning and trade. The main crops are wheat, barley, and tobacco. In some of the warmer areas, rice is cultivated.

Hazaras raise sheep, horses, and camels. Their horses are especially bred for hard work in the mountains. The wool from Hazara sheep is considered the best in Afghanistan. Camels are raised for their hides.

Social Organization

Like the Pashtuns, the Hazaras are organized by "tribes" or large clans. There are about 15 of these super-clans. The members belonging to a specific clan usually live in a certain geographic region. With so many clans, historically, there has been much inter-clan rivalry

Tribe/Clan: Examples

- **Daizinjat** clan lives in the area of Badsken, which borders on Herat and Koshak in the east. In Khila there are about 1,000 families. The Daizinjat tribe has about 30 clans. The most powerful is the Kaka clan.
- **Daikundi** consists of five tribes: Dolat Beg, Roshan Beg, Heider Beg, Gawash, and Barat. The greenest land of the Hazaras is in Daikundi, which is also called Hazarastan Kashmir. Before the Hazaras, the Tajiks lived in Daikundi.

Local and Central Governmental Authority

Because of their position as a religious and ethnic minority in Afghanistan, there has been constant tension between the local Hazara tribes/clans and the central authority of the country. The tension has boiled over a number of times into open revolution against the central authority, most notably in the early 1890s.

Ethnic Groups: Uzbek

Definition

The Uzbeks (pronounced OOZ-becks) are a Turkic-speaking ethnic group, which makes up about 6% to 8% of the population of Afghanistan. Most of them are Sunni Muslims. They are also the dominant population in the neighboring country of Uzbekistan.

http: news.hbc.co.uk hi english sialic in depth world 2001 war on terror key maps /ethnic groups.stm

Location

Uzbeks live in a population belt that stretches along the northern border of Afghanistan from Maymaneh through Mazar-e-Sharif and Konduz. This Uzbek population belt parallels Afghanistan's northern border with Turkmenistan and Uzbekistan.

Social Structure

There are two types of Uzbeks: farmers and nomads. The farmers and city dwellers are mainly businessmen and tradesmen. The nomads raise sheep, goats, and cattle.

Settlements

The basic living unit is the village. Again, there is a nomadic village and a more permanent farming/commercial village. Both villages are based on kinship.

[] The nomadic village is mobile, moving to pasturage as the seasons change.
[] The farming village is a concentrated, permanent settlement: Houses are built close together. Most houses are walled and form a compound with a central courtyard.

Importance of Uzbeks in Afghanistan

Although a smaller minority in Afghanistan, the Uzbeks have played an important role in the country. One of the most important and most controversial groups who opposed the Taliban was the forces under the control of the Uzbek leader Abdur Rashid Dostum.

[] Dostum had allied himself with the government in Kabul but later turned against it and joined with the *mujahidin* in 1992.

[] He is currently "at odds" with the present-day government of Afghanistan. http://www.afghan-info.com/Politics/Afghan Mujaheedin/Warlords.htm

Recent Afghan History

Introduction

This section emphasizes recent history of Afghanistan to show the events that led up to the Taliban taking power in the country.

http://nths.newtner.k12.il.us/academics/social/conflict/4s%20Web%20Pages/Period9/Afghanistan/afghanhome.htm — a more detailed and lengthier history of the country.

Communist Take-over

[] In 1978, during the Saur Revolution, the Afghan communist party under Muhammad Taraki overthrew the republican government. This communist party, called the People's Democratic Party of Afghanistan, started land redistribution, secularized the country and began to re-educate the people.

[] Factionalism caused the party to break down. This allowed the Soviets to invade in 1979. Eighty thousand Soviet troops invaded the country. The Soviets replaced Taraki with their puppet Babrak Kamal. http://www.afghan-info.com/Historv/Polevent. htm#saour

At this point groups of Islamic tribes/clansmen formed a loose alliance that collectively became known as the *mujahidin.* They were anti-Soviet and anti-Communist.

Soviet Domination/ Rise of *Mujahidin*

During the 10-year occupation of Afghanistan, about five million Afghanis fled the country; close to 1.5 million were killed.

[] During this period, the *mujahidin* gained prominence as the group that fiercely opposed the Soviets. They were supported by the West and by Muslim countries in their struggle against the Soviets and the puppet government that the Soviets had imposed on the country.

[] The Soviets tried a policy of reconciliation with the opposition *mujahidin* in 1987, but the seven-party *mujahidin* alliance refused to observe a cease-fire and to participate in negotiations with the Soviet puppet regime.

Continued on next page

Soviet Domination/ New Constitution

A new constitution was adopted in July of 1987, allowing the formation of other political parties beside the communists.

[] The communist party of Afghanistan said it would be willing to share power with opposition groups in a coalition government.

[] The name of the country was changed back to the Republic of Afghanistan from the Peoples Democratic Republic of Afghanistan.

Soviet Withdrawal

The Soviets withdrew their troops in 1989. The last president, Najibullah, remained in power. Civil war continued as the *mujahidin* were strongly against the government of President Najibullah.

***Mujahidin* Rule: New Government**

A number of attempts were made to integrate the *mujahidin* in the central government through advisory councils of tribal leaders.

[] Eventually Rabbani was elected president of the country.

[] In 1992, the Peshawar Agreement was ratified: A purely Islamic interim government was installed in place of Najibullah's.

New Rules for Afghan Society

[] Leadership in this new government rotated among the seven Islamic *mujahidin* groups.

[] This coalition government banned alcohol and gambling, required women to wear the veil, and separated education into male and female institutions.

Factionalism

Infighting among the various factions of the *mujahidin* continued, even though each faction had two ministerial posts in the government.

[] Eventually the alliance broke down into civil war. The most serious conflict was between Premier Hekmatyr, a Pashtun, and President Rabbanij a Tajik, together with his defense minister, Ahmad Shah Massoud.

[] The result of this armed conflict was the almost total destruction of Kabul and more important, Pakistan's support for a group of Pashtun religious students, the Taliban.

Taliban's Rise to Power

Early Advances In September of 1994, the Taliban (meaning students in Persian), an "army" of former Muslim seminarians and religious students, entered the civil war. By 1995, the Taliban had claimed victory in the town of Herat and had advanced on Kabul, the capital.

During 1995, the Taliban made extensive gains in the country, particularly in the south. In many instances, they were welcomed by the local Afghanis who were tired with and disgusted by the chaos, corruption, and instability that mujahidin factionalism had caused.

Occupation of the Capital

- By autumn of 1996, the Taliban had driven President Rabbani out of Kabul, captured the capital, and executed Najibullah. In May 1997, after a brief alliance between opposition forces and the Taliban, violent civil war again broke out.

- Fighting then continued between the United National Islamic Front (Northern Alliance), consisting of the *mujahidin* factions, and the Taliban.

Afghanistan divided: Taliban The Taliban held about 85-90% of the area of Afghanistan. http://europe.cnn.com/SPECIALS/2001/trade.center/afghan.civil.html , http:.7w\\\vl.cnn.com/WORLD/'J61()/(i4'alahan/iiidc\.lHiv - good discussion of the problems the Taliban have in uniting the country.

United Front/Northern Alliance

- The Northern Alliance held the remaining 10% to 15% of the country. As of late November 2001, the alliance had extended its control over the country, capturing Kabul.

- The alliance was strongly against the Taliban's extreme fundamentalist methods and religious philosophy and was very much opposed to Pashtun ethnic dominance, http://www.washingtonpost.com/wp-srv/nation/graphics/attack/zone 3.html

http://www.afghan-web.com/politics/parties.html - lists anti-Taliban parties/groups and their leaders.

http://www.afghan-web.com/ciriello/alliance/ - pictures of alliance forces.

Taliban: Who Are They?

Education and Training

The Taliban were originally young Afghanis many of whom went to Pakistan to study in madrassas (religious schools). These schools:

[] Focus on instruction in the Qur'an; however, a number of them teach military training and militant Islam to their students.
[] Were the source of the Taliban's interpretation of Islam and their approach to restructuring Afghanistan socially and politically. (Saudi and Pakistani intelligence services also trained these Afghan religious students.)

With their "purist" Islamic view, the Taliban were considered a group that promised a clean, uncorrupt society for Afghanis and were opposed to the factionalism and corruption of the *mujahidin,* who had plunged Afghanistan into civil war after the withdrawal of the Soviets.

Outlook

The Taliban are very provincial and backward in their outlook: The external world, much of which they considered unclean and corrupt, begins at the border of Afghanistan.

The Taliban have a close relationship to Osama bin Laden, whom they feel follows the same fundamentalist precepts they believe in. Their relationship to bin Laden was further cemented by the marriage of their leader, Mullah Omar, to one of bin Laden's daughters.

Taliban Oppression

Besides the requirement for women to be fully veiled and accompanied by a **man** in public, the Taliban had

[] Stopped all education for girls and women
[] Forbidden women to work
[] Forbidden women to wear white socks
[] Required men to grow beards
[] Stopped the Afghan national sport of Buzkash
[] Changed the name of the country to the Emirate of Afghanistan
[] Forced conscription upon the male population of Afghanistan to increase their fundamentalist army

The law of the country was the Taliban interpretation of the Sharia, Islamic religious law, with strict penalties, such as death and stoning, for infringements of the various oppressive laws imposed on the Afghan population.

Taliban Relations with Neighboring Countries

Iran

Tensions between Taliban-controlled Afghanistan and Iran were always high. The Iranians were angered by the murder of their diplomats and a journalist by the Taliban near Mazar-i-Sharif. In response to these killings, in September 1998, Iran deployed 70,000 troops to conduct military exercises near the Afghan border, but no fighting occurred between the two nations.

[] The Iranians also feel a "kinship" to the Shiite Hazaras of Afghanistan, whom the Taliban have persecuted and massacred.

[] Finally, the extreme primiliveness and barbarity of Taliban fundamentalism alarmed Iran. The Iranians were probably glad to see the Taliban removed from Afghanistan as the controlling government authority.

Pakistan

The Taliban were supported among some groups of the Pakistani population and by some Afghan refugees living in Pakistan, particularly those close to the Afghan-Pakislani border and those trained in the militant madrassas. The kind of fundamentalism preached by the Taliban is also prevalent in some sectors of Pakistani society.

Although the government of Pakistan supported the U.S. effort to root out bin Laden and his network of terrorists based in Afghanistan, Pakistan still recognized the Taliban as the legitimate government of Afghanistan long after the rest of the world denied their legitimacy.

Other Countries

Tajikistan, Uzbekistan, and Turkmenistan were all apprehensive about the Taliban, particularly their type of Islamic fundamentalism. Many of these former regions of the Soviet Union, now independent countries, are struggling to establish democratic processes for their populations. They viewed the spread of Taliban Islamic fundamentalism as a challenge to their own move toward popular sovereignty.

On the other hand, since there are minorities from these countries, particularly Uzbeks and Tajiks, living in Afghanistan, the governments of these neighboring countries had to be careful about the amount and intensity of their anti-Taliban support.

Post-9/11 Afghanistan

9/11: Activating the US War on Terrorism

September 11, 2001 is a date that saw the United States and eventually other coalition countries begin a concerted effort to actively combat international terrorism, primarily represented by Osama (Usama) bin Laden. He was eventually placed on the FBI's "Ten Most Wanted List" with a $5 million reward for information leading to his arrest.

After 9/11, the U.S. government requested that the Taliban government of Afghanistan turn over bin Laden, the international terrorist who had masterminded the terrorist attacks on the World Trade Center (WTC) and the Pentagon.

[] When the Taliban refused to extradite bin Laden, a US-led coalition was formed and after six weeks of aerial bombardments and military action on the ground the Taliban was ousted from power on November 17, 2001.

[] Many Taliban dispersed to mountain retreats in the country or fled across the border into neighboring Pakistan. Osama bin Laden is still at large and his whereabouts are not known. The coalition hunt for these remaining terrorists and for bin Laden continues. (A brief discussion of military action in Afghanistan, including Marine Corps participation, is in **Fighting Terrorism: Operation Enduring Freedom** and **Brief Military Chronology,** both later in this handbook.)

Planning for a New Afghan Government

[] In late 2001, important leaders of the anti-Taliban groups in the country and leaders from the Afghan diaspora [Afghanis living in exile abroad] met in Bonn, Germany.

[] The groups signed onto the Bonn Agreement, a plan to create a new government in Afghanistan.

[] A judicial commission was also established to rebuild the justice system according to Islamic principles, international standards, the rule of law and Afghan legal traditions.

New Government, Old Problems

Afghan Interim Authority

The beginning of the new Afghan government was the Afghan Interim Authority (AIA), which was established, with Hamid Karzai as its chairman on December 22, 2001.

International Pledges

In January 2002, to support the new Afghanistan government, sixty countries and international financial organizations present at the Tokyo Donors Conference pledged donations totaling $4.5 billion through 2006 for the reconstruction of the country.

TISA

The Afghan Interim Authority (AIA) held a nationwide grand assembly, called the Loya Jirga, in June 2002. Karzai was elected president of the Transitional Islamic State of Afghanistan (TISA) for a two-year term by the Loya Jirga. The TISA has:

[] 18 months to prepare for and hold a Loya Jirga [grand assembly] to adopt a constitution for Afghanistan.
[] 24 months in which to hold national elections.

In December 2002, the TISA marked the one-year anniversary of the downfall of the Taliban regime.

hllp://www.nulshellnotes.com/afghanistan lext.htm has a good update on Afghanistan, which is succinct and accurate.

Continued on next page

New Government, Old Problems, continued

Continued Factionalism

Factionalism is still a critical issue in Afghanistan even though there is one transitional government. The split between the ethnically dominant Pashtuns, some of whom were Taliban, and the "minority" tribes of northern Afghanistan, such as the Tajiks and the Uzbeks is a hindrance to real unit}'.

[] Besides this "tribal" rivalry, foreign involvement by groups vying for oil reserves complicates the problem of unifying the country.

[] Other countries still fear the spread of residual fundamentalism left from the Taliban and object to possible civil rights abuses in present-day Afghanistan.

[] The opium trade in Afghanistan also contributes to the factional problem.

Political Parties

Afghan political parties are not stable either. Karzai's government is a coalition of leaders from all political and ethnic groups. Various leaders in the country are planning to create new parties. Some of these political factions in the current government have formed new groups and parties, intending to participate in the 2004 elections.

Fighting Terrorism: Operation *Enduring Freedom*

Definition

Operation *Enduring Freedom* is the name given to US combat operations against terrorism in Afghanistan. *Enduring Freedom* began on October 7, 2001.

Purposes

The primary purpose for Operation *Enduring Freedom* is to seek out Osama bin Laden, who had been given sanctuary by the Taliban.

The Taliban who had controlled Afghanistan had allowed *Al Qaida* (Al Qaeda), the key terrorist group lead by Osama bin Laden, to establish training and base camps in Afghanistan. Thus another important reason for Operation *Enduring Freedom* is to root out and destroy these training and base camps and bring in-country members of Al Qaida (Al Qaeda) to justice.

A third purpose is to seek and destroy Taliban and anti-government forces operating in Afghanistan. As mentioned earlier, many Taliban members dispersed to mountain retreats when the coalition invaded Afghanistan. These residual fundamentalist militants are the focus of coalition search and destroy missions.

Another important reason is to ensure the stability that is necessary to reconstruct Afghanistan politically, economically, and socially. Without a stable environment, any new Afghan government will have a very difficult time uniting the country under one central government authority.

Continued on next page

Key Elements

Sea based: three strike aircraft carriers, each with various fighter squadrons aboard; the USS *Kitty Hawk,* which transported Army Special Operations helicopters; two amphibious groups built around the 15' and 26' MEUs, both of which later became the main US ground force in southern Afghanistan.

Diego Garcia base, which essentially housed B-1 and B-52 bombers.

Land basing in neighboring countries: Uzbekistan allowed US helicopters to operate out of Khanad; a few based in Tajikistan; Pakistan offered bases covertly out of Pasni and Jacobabad. These bases enabled the special forces assault on 17 October and the Marines' forward operation base (FOB) Rhino.

Precision Air Attack: Joint direct attack munition (JDAM) which Hew to preset coordinates based on satellite information (GPS/INS). Wind-corrected munitions (bomblet) dispenser (WCMD) first used in Afghanistan; laser-guided bombs. Electronic sensor surveillance used unmanned air vehicles.

A big plus was the fact that heavy bombers could engage multiple targets with JDAMs.

NATO Involvement

On September 12, 2001, the North Atlantic Treaty Organization (NATO) met in emergency session and, for the first time in its history, invoked Article 5 of the Washington Treaty. This article states that an attack against one NATO country is an attack against all member countries. As a result:

[] All 19 NATO Allies and the 9 countries applying tojoin NATO have provided blanket overflight rights, access to ports and bases, refueling assistance, and increased law-enforcement cooperation.

[] NATO Airborne Warning and Control System (AWACS) crews logged over 3,000 hours patrolling and protecting American skies while U.S. planes flew missions in Afghanistan.

[] Sixteen NATO allies now support Operation *Enduring Freedom* in Afghanistan and the global campaign against terrorism.

[] Fourteen allies have deployed forces in the region.

[] Nine allies are participating in combat operations in Afghanistan.

Continued on next page

NATO Involvement, continued

[] Allies and other partner countries have deployed nearly 4,000 troops to Afghanistan and have also provided 95% of the international security assistance force (ISAF), led first by the United Kingdom and now by Turkey.

[] As of August 2003, NATO took control of all security in Kabul. This is NATO's first operational commitment outside Europe.

See more details at: http://www.globalspecops.com/nalo.html

Results

Twenty-seven days after the 9/11 terrorists attacks on the World Trade Center and the Pentagon, the US joint team invaded Afghanistan and two months later freed Afghanistan from Taliban domination, together with US allies.

This was accomplished without:
[] Any US bases in-country
[] A major plan for removing the Taliban from power.

http://www.leieune.usmc.mil/2dmardiv/36/ has detailed reports and lessons-learned from Marne Corps experience in Afghanistan. Click on the link above; click on operations, then on professional forum. Then click on any of the reports individually.

Continued on next page

Key Accomplishments

[] The deepest amphibious operation in Marine Coips history - over 400 miles. Task Force 58 conducted operations in Afghanistan far from the sea. The operations showed that Marines could operate expeditionary forces at extended distances from a sea base.

[] Marines and soldiers fought battles at extremely high elevations, over 10,000 feet above sea level.

[] *Enduring Freedom* was the first time C-17's were employed in a medium threat environment to air drop supplies.

[] For the first time, headquarters and the aircrew viewed a picture taken from an unmanned vehicle simultaneously over the target.

[] Operation *Enduring Freedom* included the US Air Force's longest combat sortie on record - 44 hours. Over all, naval aircraft flew about 75% of all sorties and dropped about 33% of the bombs. The Air Force flew about 25% of the combat sorties and dropped about 75% of the bomb tonnage, including the "Daisy Cutter."

Brief Military Chronology

Introduction

This section is an abbreviated discussion of military action during Operation *Enduring Freedom*. The topics include the following:

Early Days

On October 7-8, 2001, The United States and Great Britain began strikes against Taliban air defenses, fuel sites, al Qaida (Al Qaeda) terrorist training camps and headquarters and military installations.

[] The main thrust of air assaults was the Taliban regime in Afghanistan: Concurrently crippling Taliban movement, the allies supported the Northern Alliance

[] Such precision bombing only stiffened the Taliban who did not hand over Osama bin Laden.

These targeted actions were designed to disrupt the use of Afghanistan as a terrorist base of operations and to attack the military capability of the Taliban regime. At the same time the U.S. and its allies began to drop food, medicine, and supplies to the population of Afghanistan.

Continued on next page

War in the North

Two issues moved the war effort forward in northern Afghanistan: increased bombing by the allies and the commitment of US Special Operations Forces (SOF). This same kind of commitment was to be important in the success of the Marines later in their search for residual Taliban and Al Qaida forces.

[] On November 9, Mazar-i-Sharif fell to the Northern Alliance, effectively opening a land route directly to Uzbekistan and blocking any northward movement of the Taliban from Kabul to Uzbekistan. Capturing this city also offered coalition forces a major airfield in-country.

[] The momentum of the Northern Alliance supported by the allies rolled on until the Taliban abandoned Kabul on 12 November.

[] The last Taliban stronghold, Konduz, surrendered on 26 November.

War in the South - FOB Rhino

Background

The Taliban had a stronger hold on southern Afghanistan. It was more difficult in the south to muster the local population to oppose them than in the north. The Taliban also had better security forces there.

MC Mission

The plan called for TF58 to seize and hold Rhino, while conducting raids from it. As General James N. Mattis, commander of TF58, expressed it:

"...to create chaos, denying the enemy their sense of security."

This coupled with guidance from Vice Admiral Moore, Commander NAVCENT/ MARCENT: The Marine raids would defeat the Taliban and Al Qaida forces quickly and decisively became the Marines' mission.

Elements

US Army Special Forces: Generally, these forces could initially seize an airfield or installation, but did not have the manpower to hold it. This is where Marine Corps forces came into play.

Marine Corps: Task Force (TF) 58 consisting of the
[] 15th Marine Expeditionary Unit (SOC) [MEU (SOC)]
[] 26th MEU (SOC).

Location

An airstrip that the Special Forces had raided, called FOB Rhino.

Continued on next page

War in the South - FOB Rhino, Continued

Establishing the FOB

Marines seized FOB Rhino on November 25, 2001.

[] They linked up with Navy SEALs who had earlier arrived there for target surveillance and reconnaissance.

[] Once the airfield was declared KC-130 capable, additional Marines were landed. The MEU's attack helicopters provided close air support.

TF 58: Interdiction of Rt. 1

This operation began on December 4, 2001, and included an approximately 85-mile, 19-hour motorized movement across the desert. The raid resulted in the:

[] Capture of a Taliban convoy.
[] Detection of a P-3.
[] Killing of 120 enemy personnel.

Capturing Kandahar

Background

Kandahar was a Taliban stronghold, the center of their power. The Taliban refused to surrender the city to Afghan Southern Alliance forces, and Al Qaida began to slaughter local Afghanis who were surrendering. Finally, on December 6, 2001, the city fell.

MC Action at Kandahar Airport

TF58 linked up with the Southern Alliance and U.S. Special Forces on December 13, 2001.

[] By December 14, TF58 had established a second FOB at the Kandahar airport.

[] Marines repaired the airstrip so that by December 18 the airport was considered safe for C-17s.

Further MC Operations

Looking for intelligence materials:

[] In villages west of FOB Rhino
Near Kandahar

War in the South: Khowst-Gardez Mission

Background

Although Kandahar had fallen to the alliance and coalition allies, strong pockets of resistance still remained. The local alliance leader had merely raised the alliance flag; his troops had not actually taken on the Taliban.

TF58's Mission

TF58 was tasked to seek out and destroy these pockets. Special Forces would identify the targets, and Marines would "hammer" them. Because the distance from Kandahar to the battle space was about 250 miles, a local airfield, about an hour from the battlefield, was used for close air support.

Disestablishing TFS8

During January 2002, TF58 was relieved in place by troops of the Army's 101st Airborne Division; however, when the Taliban attacked the Kabul airport, the Marines beat them back. TF58 was disestablished at the end of February 2002.

Tora Bora and Operation *Anaconda*

Al Qaida (Al Qaeda): Two Options for Escape

Al Qaida, including Osama bin Laden, and surviving Taliban leaders and forces had had two possibilities for escape from invading coalition forces:

[] They could leave Afghanistan by escaping to Pakistan or through Pakistan to countries that harbored groups sympathetic to the Al Qaida cause, such as Somalia, Indonesia, or the Philippine Islands.

[] The second possibility was to go into hiding in the mountain caves and retreats in the Tora Bora Mountains in eastern Afghanistan, near the Pakistani border.

Tora Bora

The assault on Al Qaida (Al Qaeda) in Tora Bora was conducted by Afghan troops strongly supported by U.S. firepower. Few U. S. troops were directly involved in the assaults. Afghan troops took the caves, but many of the key Taliban and Al Qaida figures there escaped.

About 300 bodies of bin Laden fighters were counted; other bodies were probably crushed beneath collapsed caves and bunkers. 150 more enemy personnel were caught trying to escape across the border into Pakistan.

Tora Bora taught Al Qaida forces that U. S. detection systems could intercept messages from radios and cell phones. Al Qaida would attempt to profit from this lesson-learned in Operation *Anaconda.*

Operation *Anaconda*

Operation *Anaconda* (in Gardez and Shah-e-Kot Valley) was mainly a U. S. forces operation. British, Canadians, and Australians also took part in the action. There were about 1,000 Afghan troops, 1,000 U. S. and some additional coalition members.

The purpose of the operation was to destroy what was considered the last pocket of Al Qaida troops, thus preventing them from launching an offense the following spring.

Continued on next page

Tora Bora and Operation Anaconda, Continued

Tactics

Limited attacks on Al Qaida (Al Qaeda) troops to force them to reveal their positions so that air strikes could be called in.

[] Apparently, Al Qaida troops attacked whenever they saw U.S. troops, regardless of the conditions of the battlefield, e.g., they would attack uphill and into fire.

[] Al Qaida also had little understanding of the sensors and sophisticated weapons they faced.

Cave Raiding: Results

Although there were some problems, in general, this "cave raiding" was worthwhile:

[] Documents and computers, with clues to future Al Qaida (Al Qaeda) operations as well as lists of Al Qaida personnel and financial accounts of the group were recovered.

[] Information about the Al Qaida's interest in weapons of mass destruction was also forthcoming.

Protecting the interim Afghan Government

In December 2001, The United Nations assembled the International Security Force (ISAF) to stabilized Kabul, the capital of Afghanistan. The bulk of the ISAF came from Muslim troops from Turkey and Jordan, with small contingents from the Canada, Western Europe and Australia and New Zealand.

Note: Much of this brief military chronology came from chapter 9 "Striking Back" of Norman Friedman's Terrorism, *Afghanistan, and America's NEW WAY OF WAR.*

Current Challenges

Introduction

Rather than try to explain the current challenges of deploying to Afghanistan, a number of "case studies" are detailed below. You should be able to get some idea of the "working" environment in country that Marines have already encountered and that you may have to face. The topics in **Current Challenges** are shown in the table below:

Local Problem: Low Intensity Civil War

"Karim Khan stands disconsolately outside the local government headquarters in the remote village of Tuksar [in Afghanistan! • *He used to run the neighbouring village, but was bundled out by a rival militia one night recently, leaving his wife and family behind as virtual prisoners."*

[] The incident is not isolated. It is not so uncommon throughout northern Afghanistan, where central government control is not as strong as it is near the Afghan capital Kabul.

[] This has led a number of experts to consider some aspects of the combat situation in Afghanistan today as low-level civil war in which local militias use the autumn, the country's traditional fighting season, to change the map of power.

Continued on next page

Feuding Warlords

Government curbs northern warlords

Feuding commanders' armies to merge as Kabul shows resolve.
By Ahmad Nahim Qadiri in Mazar-e-Sharif (ARR No. 79, 29-Oct-03J

"Afghanistan's transitional government has taken decisive action to improve security in the north of the country, dismissing key officials in Balkh province and ordering two feuding military forces merge. In a major shake-up announced by interior minister Ali Ahmad Jalali on October 26, the governor and deputy governor of Balkh province were replaced and the chief of police in the administrative centre Mazar-e-Sharif was dismissed.

"Jalali also announced that the two army corps run by rival commanders General Abdul Rashid Dostum and General Atta Mohammad are to be merged into a single force and integrated into the new Afghan national army. Disarmament of heavy and light weapons is scheduled to begin in a few weeks in Mazar-e-Sharif, supervised by the national defence ministry, the United Nations Mission in Afghanistan, UNAMA, and the British-led provincial reconstruction team, PRT.

"Analysts in Mazar-e-Sharif say the changes are designed to strengthen the hand of President Hamed Karzai's government in the north of Afghanistan, ahead of crucial decisions on a new constitution and elections planned for next year. They say the decisions are important for symbolic well as practical reasons. ... The changes come amid heightening tension in Mazar-e-Sharif, where there were fears that a ceasefire agreement between the forces of Dostum and Atta Mohammad could collapse.

"Dostum, who leads the Junbesh-e-Milli movement and has been a key player in the north for more than a decade, has been battling against Atta Mohammad for control of Mazar-e-Sharif and surrounding territory since the fall of the Taleban regime nearly two years ago.

'The two men control large forces — designated the Eighth and Seventh Corps, respectively - which formally come under Defence Minister Mohammad Qasim Fahim, who like Alia Mohammad is part of the powerful Jamiat-e-Islami faction. The turf war between the two groups has frustrated attempts by the transitional government to extend its authority to the northern provinces."

As you can see, this article points up a number of issues that in-country deployed forces may have to deal with:

[] Indigenous forces that are not united
[] Tension between the central government and locally based leaders
[] Local vice the national view on demobilization
[] Turf-war claimants

From:
http://ww.reUefweb.int%/rw^^
256dcl"0012()l"a8?Opcn] document

Continued on next page

Introduction

As a member of 3rd Bn, 6th Marines, L Co., 1st Plt., 26 MEU (SOC), these are the experiences of an infantry platoon sergeant who served in Afghanistan:

My platoon was the security element for the company. We arrived in Afghanistan on 19 Dec 2001 and stayed until 5 Feb 2002.

FOB Rhino: Insertion

[] We flew from Pasni, Pakistan, into FOB Rhino on 19 Dec 2001. We went from 90-degree heat to 40-degree chill in 3 hours. What an adjustment! After the first couple nights we became accustomed to the drastic temperature change.

[] The biggest challenge was the feeling that we were isolated. Since 1/1 was securing the outer perimeter we only had to man the interior posts. After 2 days we were flown by C-130 to Kandahar.

Kandahar

Purpose

Our purpose at Kandahar was to augment security of the airfield:

[] We didn't have much interaction with the civilians.
[] The civilians we did meet seemed very pleasant and not hostile in any way.

Our translator told us the people were very glad to have security and stability in their lives for once.

Climate

The temperature in January ranged from 50° F. in the day to 15 °F. at night. You can expect a 20-30 degree temperature change when the sun sets.

It was dry except for two rainy days. When it rains, the sand and hard packed dirt become ankle-deep mud. The mud is slippery and it stinks!

Kandahar (Continued)

Keeping Warm

Marines will find ways to stay warm no matter what. Here are a few ways we kept warm.

[] While on watch Marines would lay their poncho liners over their laps.
[] Hot wets were prepared before nightfall and served out to the Marines.

These little things plus the gear we brought with us made our nights passable.

Keeping Clean

Since we didn't have showers, and water was at a premium, we had to take canteen cup baths, better known as birdie baths.

[] We would heal up a canteen cup full of water and wash ourselves head to toe.

[] With that same canteen cup of water we would then wash our socks and tie them to our packs to dry in the daylight.

After a few weeks we received wash buckets to wash our clothing. Teaching Marines how to wash clothes by hand and with a washboard was unique!

Making Life Bearable

We had packed enough tobacco products to last 30 days. Since the army took nearly a month longer to get there, we made arrangements with some air force personnel to bring some tobacco in to us once in a while.

Mail didn't arrive in the country until the week before we departed. Priorities were water, food, equipment, ammo, fuel, and then mail. However, the MEU did set up three to four phones for us to use. We would have to wait for hours to make a call, but it was worth it.

Coffee was made by pulling a clean sock tip around the top of a canteen cup; put the coffee grounds on the sock, then pour hot water over the grounds. Coffee made a huge difference!

Continued on next page

Kandahar (Continued)

Dangers

While in Kandahar there were a few things we, Marines, always kept in the forefront of our minds: mines, mines, and more mines.

[] Not only did we have to watch for Soviet mines, we had to be careful of our own cluster bomblels. Wherever we saw some in the area, we had our Marines take a look at Ihem so Ihey would know what they looked like.

[] Unexploded ordinance was everywhere. EOD was busy with the bigger stuff. All our Marines could do was mark near where the ordinance were and avoid the area.

Important!

Some precautions that paid off:

[] Gear inspections and gear accountability (No one lost any gear!)
[] Uniformity within the company
[] Entrenching was continuous. We made bunkers not just fighting positions.
[] Weapons training (not one ND while we were there!)
[] Training the NCOs (Never did I catch a Marine breaking light discipline or falling asleep on watch) The NCOs ensured their Marines were alert and prepared!

Raids near Khowst

Purpose

Our purpose near Khowst was to conduct a raid on cave complexes located near there. The raid was supposed to last under 10 hours from insertion to extraction; however, it lasted 10 days.

[] In those 10 days we gathered much of our intelligence information.
[] We also proved that Marine units could and did survive, fight, and prevail without resupply.

Climate

Khowst is located in northeast Afghanistan, near the border with Pakistan. The climate in the mountains at night was extremely cold, in the teens to single digits. During the day the temperature was 40°F-50°F. There were trees and scrub brush in the area. It didn't rain during the time we were up there.

Continued on next page

Raids near Khowst (Continued)

Terrain

The terrain is mountainous. Physical preparation and education is needed for operating in a mountainous environment.

[] Vehicle movement is restricted to the *Wadis* and roads,
[] There are many old bunkers, fighting positions and caves throughout the mountains.
[] There were trees and scrub brush in the area.

Enemy Tendencies

The enemy made bivouac sites opposite their caves. The cave locations were mostly in the *Wadis* where they were extremely difficult to see or bomb.

[] We found equipment slashed in burial sites, buried outside villages, and just left plainly in the open.
[] The enemy signaled at night from mountaintop to mountaintop with white light.
[] There was also evidence that they had cell phone/satellite phone capability.

As soon as we landed, we found evidence thai the enemy had hastily withdrawn into Pakistan. After a few days we had a helicopter arrive to take detainees to Kandahar. Some started making their way back over the mountains.

Interacting with the Local Population

Some of the people we took as detainees were absolutely scared of us. Based on our physical size and appearance, they would become almost hysterical, but with a little show of kindness on our part, they willingly told us many tidbits of information: All we had to do was give them a Koran and let them know when it was prayer time. All of a sudden we discovered that one of them could speak English.

[] These people knew which group controlled what area; they would claim allegiance to both so thai they could have unfettered access to both areas.

[] The people seemed to side with whoever was more powerful.

[] They have little national pride; they did have pride in their tribes!

In a vehicle we captured going down the ***Wadi*** we found a Taliban flag and a Northern Alliance flag. In the same vehicle we also found a GPS.

Continued on next page

Raids near Khowst (Continued)

Living off the Land

Because we had no re-supply, we had to live off the land (a 10 hour-raid means you carry very little gear).

[] We seized an abandoned terror training facility and found livestock and grain.

[] We soon began making corn bread, pop corn, and other types of food. After a couple days we had chicken, lamb, and steak.

[] We found water in cisterns and in natural springs in the *Wadi.* We purified the water in 5-gallon jugs.

Small Unit Operations

In the mountains we had great success, operating at squad size and less. There were always 2 patrols out day and night. They would range 3-4 km and stay at altitude. This helped us gain vital information of enemy movements and discover weapons caches.

We constantly used binoculars and NVGs with 3x Mag for observation. This timely warning and information meant we could capture many people who may have otherwise become combatants.

http://www.lejeune.usmc.mil/2dmardiv/36/ has detailed reports and lessons-learned from Maine Corps experience in Afghanistan. Click on the link above; click on operations, then on professional forum. Then click on any of the reports individually.

Continued on next page

Broad Tactical Environment

By Jim Garamone

American Forces Press Service
WASHINGTON, June 4, 2003 - U.S. and Italian forces launched an offensive June 2 against anti-coalition forces in the Shahi Kovvt section of the Paktia province in Afghanistan, Combined Joint Task Force 180 officials said today.

A brigade-sized unit launched Operation Dragon Fury to prevent the re-emergence of terrorism, to deny sanctuary to terrorists and to protect nongovernmental organizations and coalition forces from enemy attack.

Army Col. Rod Davis, spokesman for the command, said the ground assault section set up blocking positions and air assault troops drove anti-coalition forces into the trap. A total of 21 men were taken under control by coalition forces.

The U.S. forces came out of Kandahar, the Italian forces from Khowst. Paktia province is located in eastern Afghanistan near the border with Pakistan. The province was the scene of tough fighting in 2001 and was raided again last year.

Roughly 20 helicopters launched the air assault. Close-air support was standing by in case the coalition troops needed the assistance, Davis said. He said the Italian forces searched more than 300 vehicles and inspected more than 800 people.

There were no coalition casualties.

American forces came from the 82nd Airborne units in the area. Special operations personnel also took part.

From: http://www.globalspecops.com/oef2a.html This article shows the "larger environment" in which Marines like those in the earlier case study might be involved in.

Rebuilding Afghanistan

Armed Forces

The United States is helping the Afghan government to build a national army. In 2002, US soldiers helped train 1,600 Afghan soldiers for this new army.

http://www.elobalspecops.com/rebuilding.html is good site that summarizes US efforts to rebuild Afghanistan.

Assistance to Women

Although the extreme persecution of women inflicted by the Taliban has ceased, women still have a difficult lime in presenl-day Afghanistan.

[] The U.S. has helped the Afghans to establish women's resource centers.

[] Over one million dollars has been used for training women in business and for educating girls.

Transportation Infrastructure

The United States has committed $800 million to reconstruct the main transportation route between Kabul, Kandahar, and Herat. The project, underway since 2002, is scheduled for completion by 2004.

The United States is helping the Afghan government to repair and reconstruct bridges and 4,000 km of secondary and tertiary roads. The United States is also helping to winterize a major tunnel that is a key link to northern Afghanistan as well as participating in revitalizing over 6,000 water wells, canals, dams, and water systems in the country.

Child Welfare

[] The United States has funded an immunization program for Afghan children. Since April 2002, 4.3 million Afghan children have been immunized against measles.

[] America's Fund for Afghan Children has raised over $11.4 million to help Afghan children. Moneys from this fund have been used to purchase school supplies and books as well as teacher kits. Funds have been used to build playgrounds. Winter relief and health kits for children have also been supplied from this fund.

Continued on next page

Rebuilding Afghanistan, continued

Refugee Welfare

More than two million refugees and displaced persons have returned home to Afghanistan since the overthrow of the Taliban.

[] The United States has donated funds and assistance to help resettle these refugees and to provide them with food and basic needs, as well as shelter kits to the neediest.

[] An Afghan Conservation Corps has been established, giving Afghan refugees who have returned to the country and demobilized fighters employment and job training.

Health

[] The central disease surveillance system for the eradication of polio was re-established, and a national curriculum for midwives was revised.

[] Health facilities, including ov er 28 clinics and hospitals, have been rebuilt and refurbished. A national plan has been drawn up to determine the location and type of construction for an additional 600 health care facilities.

Economic and Agricultural Activities

The Afghan government has an ambitious business enterprise policy. The aim is to restore full central government authority over the nation's currency and fiscal system. The United Slates is assisting the Afghans in developing banking, trade, and fiscal and business reforms. More privatization of stale owned banks and public sector organizations is also sought.

Seventy percent of the Afghan population is engaged in agriculture. The United States has funded assistance to Afghan farmers aimed at improving crop production. In the spring of 2002, the United States donated 7,000 tons of seed, fertilizer and technical support to over 40,000 farmers. During 2002, crop production increased 80% over 2001 production.

http://vvvvvv.developmentgateway.org/node/134111/ is an excellent site on reconstruction in Afghanistan, showing what various coalition members are doing to help the country.

Islam: Basic Ideas

Introduction

You have probably heard the name *Islam* associated with Afghanistan. Most people in Afghanistan and in surrounding countries believe in Islam. The Taliban and Osama bin Laden believe in a radical form of Islam. In this section, you will learn about Islam, its main ideas, customs, and practices.

(http://thetruereligion.org/intro.htm - - a short introduction to Islam)

Definition of Islam

Islam means submission to the will of God in the Arabic language. The word *Islam* has the same root as the Arabic word *salaam,* which means peace. Muslims around the world often greet each other with the phrase "Salaam" in much the same manner as Jews say, "Shalom," as a greeting. Both salaam and shalom come from the same common language root, the three-letter root - - *s-l-m.*

(http://media.isnet.org/off/Islam/basics/index.html) - discussion of various "doctrinal" aspects of Islam by noted theologians.

Allah

Islam is not named after a person like Christianity, which was named after Jesus Christ, Buddhism after Buddha, and Confucianism after Confucius. The central focus of Islam is always God.

The basic tenet of Islam is that you must submit to Allah (the Arabic word for God) and live according to His divinely inspired law.

The Precept

[] The key truth that Allah has revealed to mankind is that the only divine and worshipful "being" is Allah, the almighty God; thus all human beings should submit to and worship Allah.

[] Allah has 99 names, among them, The Gracious, The Merciful, The Beneficent, The Creator, The All-Knowing, The All-Wise, The Lord of the Universe, The First, The Last, and others.

No matter what "sect" of Islam a person belongs to, he believes in this important religious belief that is the common thread of Islam.

Arabs and Islam

Muslims (Moslems)

Now that you know that the religion is called Islam and the name of the God of Islam is Allah, what does the word *Muslim* mean? The word has the same s-l-m root as the words *Islam* and *salaam.* A Muslim is a person who submits to the will of Allah, regardless of race, nationality, or ethnic background. A Muslim has to submit completely and obediently to Allah and live according to Allah's message.

[] Muslims worship Allah alone and must not worship any person, place or thing other than Allah. They believe that Allah is the God for the Christians, the Jews, the Muslims, the Buddhists, the Hindus, and even atheists.

[] Orally repeating the basic belief of a Muslim found in the "motto" of Islam: *"Laa Elaaha illa/laah"* which means, "There is no god but Allah," is the way a Muslim professes that he belongs to Islam.

Importance of Arabic in Islam

The "motto" above, *"Laa Elaaha illatlaah"* is in Arabic. Even though the people of Afghanistan speak a number of different non-Arabic languages, the language of Islam is Arabic, and this is the language used in "religious services." (Historically, Islam began among the Arabs of what is now Saudi Arabia.)

[] The prayers, sayings, and many blessings of Islam are all spoken in Arabic, just as the holy book of Islam, the Qur'an (Koran) is written in Arabic.

[] This common religious language of Islam is one of the ties that bind all Muslims together.

Continued on next page

Arabs and Non-Arab Muslims

Because of the importance of the Arabic language in Islam, many people assume that most Muslims are Arabs. This is not the case: About 80% of all Muslims are **not** Arabs.

[] In fact there are more Muslims in Indonesia than in the Arab Middle East.

[] Besides Indonesia, Muslims make up the majority in such non-Arabic countries as Turkey, Pakistan, Uzbekistan and other central Asian countries of the former Soviet Union.

[] There is a large minority in China too.

[] In Europe, Albania is Muslim. Bulgaria, Bosnia, and Macedonia have large Muslim populations. Elsewhere in Europe, there are immigrant communities of Muslims from Africa, Turkey, and Asia in France, Britain, and Germany.

[] In the Americas Muslims have increased in recent years, both from conversions and immigration; 20% of the population of Suriname in South America is Muslim. Today there are about five million Muslims in America.

Muhammad (Mohammed)

Importance

Muhammad is ... the Messenger of God
and the last of the prophets ... (Qur'an, 33:40)

This one phrase from the Qur'an shows Muhammad's place in Islam and in world religion. Muslims believe that Muhammad is the prophet of Islam. His message is Islam. The revelation of Islam that Muhammad received is the *Qur'an* (Koran).

Muhammad and Other Religions

In terms of other world religions, Muhammad is the last prophet of God to mankind.

[] He is the last messenger of God.
[] His message applies to all mankind, regardless of their beliefs.
[] Muhammad is the successor to Moses, Jacob, Isaac, Abraham, and Jesus.

Muhammad is the prophet among all previous prophets and messengers.

[] He reinterpreted and corrected the message of Allah and gave mankind the true word from Allah. In a sense, Muslims consider that he "cleansed" the message.

[] In spite of his importance in Islam and for the world, Muslims do not worship him or ascribe any divine qualities to the prophet.

Muhammad, His Calling

Muhammad was born in Mecca (Makkah) in what is now Saudi Arabia, in the year 570 CE. His father died before he was born. Soon after his birth, his mother also died, so Muhammad was raised by his uncle.

Because of his trustworthiness and sincerity he was often asked to arbitrate disputes. He has been described as calm and meditative. Muhammad was very religious. It was his habil to meditate in a cave near Mecca.

At the age of 40, during one of his retreats, Muhammad received his first revelation from Allah through the Angel Gabriel. This revelation, which he continued to receive for twenty-three years, is known as the *Qur'an* (Koran).

Continued on next page

Muhammad (Mohammed), Continued

Spreading Islam

As Muhammad began to preach the truth that Allah had revealed to him, he and his small group of followers suffered persecution.

[] This became so strong that in the year 622 CE Muhammad and his followers emigrated from Mecca.

[] This event, called the *Hijra* ('migration'), when they traveled about 260 miles to the north to the city of Medina, marks the beginning of the Muslim calendar.

[] After several years, Muhammad and his followers returned to Mecca, where they made peace with their enemies and established Islam.

[] Before Muhammad died at the age of 63, most of Arabia was Muslim, and within a century of his death, Islam had spread to Spain in the West and as far east as China.

Five Pillars of Islam

Introduction

The five tenets are the foundation for all Islamic belief. They are the cornerstones of Islam. (http://www.usc.edu/dept/MSA/fundamentals/pillars/ - an introduction to the five pillars.)

Commitment

Called the Ash-Shahaadah, this is the "credo" of Islam:

"I bear witness thai there is none worthy of worship except Allah and I bear witness that Muhammad is His servant and messenger."

All Muslims subscribe to and repeat this belief.

Prayer

Called As-Salaat in Arabic, this is the requirement that adult Muslims pray at five specific times during a twenty-four hour day to Allah.

[] This is a ritual prayer requiring Muslims to say specific prayers and to coordinate hand and arm movements, eventually culminating in complete prostration or bowing down with the hands on the floor to show complete submission to Allah.

[] While performing As-Salaat, Muslims face in the direction of Mecca.

Alms

Known as the Az-Zakaat in Arabic, this means that Muslims must contribute to the support of those less fortunate Muslims—support of the poor. Muslims are encouraged to pay their Zakaat during the holy month of Ramadan.

Continued on next page

Five Pillars of Islam, continued

Fasting

In Arabic this is called As-Sawam. It means that during the month of Ramadan, adult Muslims who are in good health

[] Cannot eat, drink, or smoke from sunrise until sunset every day of this holy month.

[] Must abstain from sexual relations.

The intention of fasting is to worship Allah. Ramadan is a sacred month because it commemorates the time when Allah first revealed the Qur'an to Muhammad. The time for Ramadan changes each year because it is based on a lunar calendar. Ramadan comes 11 days earlier each year.

Pilgrimage

Called Al- Hajj (Hadsh), this pilgrimage requires a Muslim in good health to travel to Mecca in Saudi Arabia at least once in a lifetime.

[] Specific preparation is necessary for the Hajj; for example there are rules about shaving and cutting hair and what to wear.

[] One of the high points of the Hajj is the *umrah,* which means performing a ritual walk seven times around the sacred stone of Islam called the Kab'ah.

[] The Hajj commemorates the sacrifices and faith of Abraham, his second wife, Hagar, and their son, Ishmael.

According to the Council on Islamic Education, it is the largest, regularly scheduled international gathering on Earth.

Al Qur'an (Koran) and Sunnah

Definition

Al Qur'an (the Qur'an) is the revealed word of Islam, the holy book of the Muslims.

[] Muhammad received the message in bits and pieces from the Angel Gabriel during a period of 23 years.
[] The word *Al Qur'an* means the recitation. During Muhammad's lifetime it was recited publicly.
[] The original language of the Qur'an is Arabic.

It is important to remember that Muslims believe the Qur'an is **the** word of Allah, untouched and uninterpreted by human beings. As Allah's final message to mankind, it is the holy book that supercedes all others: the *Old Testament,* the *Gospels,* etc.

Form and Content

The Qur'an is a collection of surahs or chapters, many of which you might call verses.

[] Muhammad received the surahs from Allah. Scribes selected by Muhammad usually wrote down these verses. Sometimes they wrote on wood, trees, parchment, and even stones. Many followers also memorized the Qur' an by heart.

[] In later years, the Qur'an was recopied and refined; accents and markings for reading were added.

Interpretation

[] The Sunnah is the spoken word and acts of Muhammad. The Sunnah, Muhammad's actual words and deeds, uses the life of Muhammad to explain and expand on the verses and teachings of the Qur'an so that Islam can eventually become a world religion.

[] A second form of interpretation is called hadith. This is narration about the life of the prophet or what he approved - as opposed to his life itself, which is the Sunnah.

The Qur'an, the Sunnah, and the ahadith (plural of hadith) are the basic body of Islamic religious doctrine for all Muslims. (http://www.afghan-network.net/Islam/) (http://www.usc.edu/dept/MSA/)

Jihad (Djihad)

Definition

The word *jihad* means straggle in Arabic. This means a straggle between the forces of Allah (good) and the forces of evil —an eternal struggle.

You have probably heard the word *jihad* in reference to Osama bin Laden and the Taliban. They consider the terrorism that they perpetuate a jihad. Actually, the word *jihad* has a couple of interpretations.

Self-defense

Islam allows fighting in self-defense and in defense of religion. There are rules of combat including prohibitions against harming civilians and against destroying crops, trees and livestock.

If good Muslims do not risk their lives in the cause of Islam, injustice would triumph. The Qur'an says:

> " *Fight in the cause of God against those who fight you, but do not transgress limits. God does not love transgressors.*" (2.190)

War is the last resort and is subject to the precepts of Islamic sacred law.

Inner Struggle

The second and more personal meaning of jihad is an inner straggle. Each of us, according to Islam, is constantly fighting a "war" against self-centered desire, and egotism, with the final goal of attaining inner peace.

Islamic Schools (Sects)

Religious vice Secular Law: Western View

The United States and most other Western countries make a clear distinction between the rule of law and religion.

[] Religion is a separate personal issue that does **not** determine law.

[] A legal case may make use of religion, but religion is not the determining factor in most legal cases.

[] Simply put, a religious denomination, such as Methodism, Catholicism, or Judaism, doesn't have the power to rule on a case of adultery in a civil court or on a case of fraud in a criminal court.

The United States has a secular system of law; religion plays no direct part in such a system.

Islamic View

If you look at Islam, the perspective is different. Islamic law consists of guidelines and rules that determine all aspects of a Muslim's life from how to perform ritual prayer to conducting business transactions.

[] It also includes crimes and the appropriate punishments for each.

[] These Islamic laws are based primarily on the Qur'an and are called the Sharia.

[] Muslims must apply these precepts handed down to Muhammad in the sixth century to present-day situations.

Interpretation and Re-interpretation

This kind of religious legal system, particularly how to interpret the Qur'an and the ahadith, has resulted in a certain degree of divisiveness in Islam, just „ varied interpretations of Christian doctrine have caused Christianity to split into various denominations. The two main sects of Islam, which are the schools of Islam, are the

[] Sunni
[] Shi'ites (Shi'ia)

Continued on next page

Islamic Schools (Sects), Continued

Shi'ites (Shi'ia)

The basic divisiveness in Islam is based on an historical incident:

[] After Muhammad died, there was confusion about who should succeed him as the leader of Islam. One group felt that the successor was Imam Ali who had been appointed by Muhammad according to Allah's decree. Ali was a relative of Muhammad and the first to accept Islam.

[] This group, mostly from the household of the prophet, was a minority and became known as the Shi'ites. They also believe that the leader of Islam must be endowed with grace and benevolence and should be infallible.

Sunnis

The majority group, called Sunni Muslims, believe that:

[] Muhammad did not choose a specific successor and probably assumed that after his death, Muslims would find their own leader.

[] The prophet did not tell his followers how they ought to select their leaders or what qualifications their leaders should have.

The Sunnis chose a leader, later called the caliph, from outside Muhammad's household. The Shi'ites refused to accept the Sunnis' choice and split off from this main group.

Results of the Division

Because of this succession split, variations in Islamic doctrine, law, and practice have developed between these two main groups over the centuries.

[] Sunni Muslims predominate in Saudi Arabia, Afghanistan, Pakistan, Turkey, and Indonesia.

[] Shi'ites are the majority in Iran and southern Iraq.

Sunnis make up about 83 percent of Muslims, according to the *Encyclopedia Britannica;* Shi'ites, about 16 percent; and a few other small groups, the remainder.

Osama (Usama) bin Laden and Islam

Bin Laden's Religious Background

Bin Laden's brand of Islam is based on extremely conservative Islam called Wahhabism, which was espoused by Muhammad bin Abd al-Wahhab (1703-87) He was the founder of the sect and the co-founder of Saudi Arabia. The Wahhabis and the royal Saud family have long intermarried. The Saudis have adopted many of the conservative precepts of Wahhabism.

Wahhabism: The Way of a Muslim Life

Wahhabi Islam is a total religious system. It has an answer for every question about a Muslim's life including social, legal and spiritual aspects. Wahhabi Islam is ascetic.

[] Men should wear short robes and even avoid the black cords used on turbans and headgear.
[] Mosques should have no decoration.
[] Drinking alcohol is forbidden.

Punishment is based exclusively on the Qur'an, for example:

[] The right hand should be amputated for theft.
[] Adulterers should be stoned to death.
[] Murder and sexual deviation are punishable by beheading. To this day Saudi Arabia uses these punishments, especially beheading for capital crimes.

Bin Laden and the Taliban

The kind of radically, conservative state that the Taliban had developed in Afghanistan had roots in Wahhabism, which inspires and feeds on the Taliban's fundamentalist religiosity.

[] For Islamic fundamentalists, the Taliban have created what approaches a pure, ideal society based on Islamic religion and law.
[] This society fits bin Laden's mold too.

The Taliban allowed bin Laden to remain in Afghanistan as a guest. They denied consistently that he had anything to do with the terrorist acts of 9/11 and refused to extradite him as requested by the US government. They continuously claimed that they did not know of his whereabouts.

Continued on next page

Osama (Usama) bin Laden and Islam, Continued

Focus on Expansion and Terrorism

[] While bin Laden follows the conservative Wahhabi tenets, what he most fervently supports is that his brand of Islamic faith must expand. (To support this kind of expansion, Saudi Arabia supported the Afghan *mujahidin* when they were fighting the Communists. So did Osama bin Laden.)
http://www.newyorker.com/FROM_THE_ARCHIVE/ARCHIVES/7010924 IV_archi\ c() - good article on bin Laden, his life and religion

[] Now bin Laden regards the struggle as a war between two civilizations: His Islam and the non-Muslim civilization, specifically the United States. Toward winning this struggle, bin Laden has committed his forces to terrorism.

http://www.brookings.Org/fp/proiects/terronsm/faqs.htm#qb4 - a set of informative questions about terrorism.

Bin Laden-sponsored Terrorism

[] Besides the infamous 9/11 terrorist attacks against the Pentagon and the twin towers, bin Laden allegedly sponsored the attacks on American embassies in Africa.

[] U.S. and Russian officials arc also concerned that he is financing Chechen rebel operations out of Dagestan, a former Soviet republic in Asia.

A primary concern is that bin Laden's forces are attempting to acquire or build chemical and/or biological weapons to use in their terrorist attacks. In the case of the Chechen rebels, he may be trying to help them acquire radiological dispersal devices.

Continued on next page

Osama (Usama) bin Laden and Islam, Continued

Al Qaida (Al Qaeda)

Al Qaida (Al Qaeda) is the name of the network of Islamic extremists that bin Laden has at his command to carry out his radical Islamic terrorism.

[] Al Qaida consists of a group of about 3,000 commanders. Troops of Afghanis and Pakistanis number 200,000+.

[] There are a number of terrorist centers, with numerous cells, in North America, Yemen, Saudi Arabia, Albania, Kosovo, Algeria, Chechenya, all the former Soviet republics of Central Asia, the Philippines, Egypt, Ethiopia and Somalia.

Bin Laden's "Armv"

Al Qaida is bin Laden's personal "army." Bin Laden could probably rally an army of over 100,000+ men around the world, excluding those currently "in service" in Pakistan and Afghanistan.

[] Many members of his overseas forces are ready for both sustained and one-time operations.

[] In the interim, the members of these terrorist forces return to normal "civilian" life in host countries. They are also available for terrorist operations on their own countries.

Funding for Al Qaida (Al Qaeda) Terrorism

The primary sources for funding Al Qaida-sponsored terrorism have been:

rj -p[illegible]f drugs, primarily opium.

[] Contributions from Saudi Arabia through various banks and from various Saudi businessmen.

[] What might be called tribute money to Al Qaida from countries that do not want the organization to establish cells in their own countries.

[] Funds from bin Laden's personal fortune.

Continued on next page

Osama (Usama) bin Laden and Islam, Continued

Bin Laden and Saudi Arabia

Even though the country where he grew up, Saudi Arabia, is extremely conservative, bin Laden finds that it is not conservative enough. He thinks the following reasons, which are contrary to his purist concept of Islam, are grounds for overthrowing the Saudi royal family:

[] Alliance of the ruling Saudi family with the West

[] Saudi Arabia's dependence on American and other foreign troops who came to the country to defend it during the Gulf War

[] Corruption of the Saudi regime

Appendix A: Short History of Terrorist Activity

Timeline The Activities of the Al Qaida

The Al Qaida has a long history of terrorist activity, which dates back to the early 1990s. Many of their earlier operations failed to achieve the desired effect; however, experience has enabled the organization to develop a more sophisticated approach.

- [] Suspected in the bombing of night club in Bali
- [] Suspected in September 11 plot to hijack planes and destroy the World Trade Center, Pentagon, and other unspecified targets (2001)
- [] Plotted an unsuccessful effort to attack U.S. and Israel tourists during millennial celebrations (2000)
- [] Implicated in bombings of U.S. embassies in Nairobi, Kenya and Dar es Salaam, Tanzania that killed over 300 (1998)
- [] Accused by the U.S. of bomb attacks on American military personnel in Dhahran, Saudi Arabia (1996)
- [] Linked to unsuccessful plot to assassinate President Clinton in the Philippines (1995)
- [] Linked to unsuccessful plot to bomb multiple U.S. trans-Pacific flights (1995)
- [] Linked to unsuccessful plot to bomb U.S. and Israeli embassies in Manila (1994)
- [] Linked to unsuccessful plot to assassinate Pope John Paul II in Manila (1994)
- [] Claims to have downed U.S. helicopters and killed U.S. servicemen in Somalia (1993)
- [] Attempted bombing of U.S. troops in Aden, Yemen (1992)

Source: U.S. State Department

Appendix B: International Associations

Top Commanders

Al Qaeda is an international association of allied groups operating in many parts of the world. Its top commanders are

[] AYMAN AL ZAWAHRI, head of the Egyptian Jihad Islami, who is Bin Laden's senior deputy and heir apparent

[] JUMMA MAMANGANI, an Uzbek, who was recently appointed Al Qaeda chief of operations. He is former commander of the Moslem Army for the Liberation of Kyrgystan. Three key Afghan training camps, at Jalalabad, Farmada and Daronta, are under his command.

[] FATEH KAMEL, who leads the most militant cells of the extremist Algerian GIA. In the name of Al Qaida, he controls terrorist cells in the United States, Canada and Algeria.

[] MUHAMMED ATIF aka SUBHI ABU-SITTAH, who is nominal chief of staff of the network and its brightest military brain. He comes from the Egyptian Jihad Islami

[] IMAD MUGHNIYEH, the former Lebanese Hizballah hostage-taker and bomber, who is in charge of the combined terrorist campaign around the Middle East, the Persian Gulf, Europe and Israel.

Throughout the world

On top of the 3,500 hard core commanders and 110,000 Afghan and Pakistani troops, Al Qaeda retains another 6400 commanders in 12 centers: North America, Yemen, Saudi Arabia, Albania, Kosovo, Algeria, Chechenya, Tadjikistan and all the former Soviet republics of Central Asia, the Philippines, Egypt, Ethiopia and Somalia.

The numbers Bin Laden can muster differs from place to place. In North America, together with his closest ally, the Egyptian Jihad Islami, some 2,500 hard core fighting men; in Yemen, where his family originated before migrating to Saudi Arabia - 2000 directly. But Bin Laden has a special relationship with the commanders of the 20,000-man strong irregular "Moslem Liberation Army, which hold a monopoly of the arms trade of the Arabian Peninsula, the Horn of Africa, the Red Sea and East Africa. Its reach sometimes goes as far as Iran.

Bin Laden often serves as the MLA's clearing bank and ready bankroll for arms deal.

Continued on next page

Appendix B: International Associations, Continued

Saudi Arabia

In Saudi Arabia, where the privileged and affluent bin Laden clan lives, Osama commands some 200-250 hard-core commanders, but there are many more potential partisans among the disaffected tribes in the central and eastern provinces, especially the Nejd, as well as in the Saudi armed forces and national guard.

An intensive Al Qaeda recruitment effort in those Saudi forces could cause their collapse and drop in Bin Laden's lap their arsenals, with some of the most sophisticated hardware in use today.

Overseas Legions

According to conservative estimates, the millionaire-terrorist could most probably rally around the world roughly the same number of fighting men as those flocking to his flag in Afghanistan and Pakistan, namely an army of over 200,000 men. Many members of his overseas legions are available for both sustained and for one-time operations. In between, they simply go back to their normal pursuits and their homes in host countries. They are also available for terrorist operations on their home ground.

Appendix C: Other References

Resources

Besides the websites and URLs cited in the body of this handbook, the following resources will help you better understand Afghan culture and people:

[] *Encyclopedia of World Cultures* (particularly volume III), edited by Paul Hockings. Boston: G.K. Hall and Co., 1992. Excellent summaries of the different tribal and ethnic groups of Afghanistan and neighboring countries.

[] *The Europa World Year Book 2001* (volume I). London: Europa Publications, 2001. An excellent in-depth survey of Afghanistan.

[] Friedman, Norman. *Terrorism, Afghanistan, and America's NEW WAY OF WAR*. Annapolis: Naval Institute Press, 2003.

[] Hoffman, Frank G. (LtCol, USMCR, Ret.). "The U. S. Marine Corps in Review." *Proceedings,* May 2002: Vol. 128/5/, 84-90.

[] Hoffman, Frank G. (LtCol, USMCR, Ret.). "The U. S. Marine Corps in *Review." Proceedings,* May 2003: Vol. 129/5/, 96-102.

[] Marsden, Peter. *The Taliban: War, religion and the new order in Afghanistan.* New York: Zed Books Ltd., 1999. In-depth study of the Taliban including their origins and type of leadership, their religious "creed," implementation of religious beliefs, etc.

[] O'Balance, Edgar. *Afghan Wars 1839-1992: What Britain Gave Up and the Soviet Union Lost.* New York, Brassey's, 1993. Good account of the major conflicts in Afghanistan over the years: the Anglo Wars, the Soviet period, and civil wars.

[] Rubin, Burnett R. *The Search for Peace in Afghanistan: From Buffer State to Failed State.* New Haven: Yale University Press, 1995.

Made in the USA
Monee, IL
05 November 2026

67808330R00105